What people ar~

Expanc

In *Expanding Reality*, neuroscientist Mario Beauregard presents the impressive body of scientific evidence that our familiar materialistic model of the world is locked into a sadly outdated 19th century philosophy. Whether or not you agree with all his conclusions, this courageous book will challenge you to rethink how we see ourselves and our world – and may help usher in a new scientific revolution.

**Bruce Greyson MD**, Carlson Professor Emeritus of Psychiatry & Neurobehavioral Sciences Division of Perceptual Studies University of Virginia Health System

As Dr. Mario Beauregard clearly and convincingly presents in *Expanding Reality*, science is experiencing another paradigm change whose revolutionary implications may be immense.

**Gary E. Schwartz PhD**, Professor of Psychology, Medicine, Neurology, Psychiatry and Surgery; Director, Laboratory for Advances in Consciousness and Health, University of Arizona

Science seeks to understand the nature of reality. But science rests upon many assumptions, some of which are incorrect and have caused science to derail. To help put science back on track, Mario Beauregard provides a clear and comprehensive survey of "postmateralistic" evidence and concepts. Read *Expanding Reality* to gain a thrilling glimpse of what is likely to become the foundations of 22nd century science.

**Dean Radin PhD**, Chief Scientist, Institute of Noetic Sciences, Associated Distinguished Professor, California Institute of Integral Studies, author of *Supernormal* and *Real Magic*

In *Expanding Reality*, Dr. Mario Beauregard lays the foundation for the next great scientific revolution, the revolution beyond scientific materialism. The reader is guided through the comprehensive scientific evidence for postmaterialism in a clear and digestible way, emerging enlightened by the science of consciousness. *Expanding Reality* breaks the chains of an outdated paradigm to usher in a new world in which we understand our interconnection and embrace our divine nature. A must-read for anyone yearning for an evolution in human consciousness.

**Natalie Leigh Dyer PhD**, Research Scientist at Kripalu Center for Yoga & Health and at University Hospitals Connor Integrative Health Network

In this very interesting and important book the overwhelming clinical, empirical and statistical evidence is shown that consciousness must be fundamental, and that conscious experiences cannot be fully understood due to the current widely accepted materialist paradigm in Western science. There is no scientific method whatsoever to objectify nor prove the content of our subjective consciousness. Unfortunately, a significant part of society remains under the influence of a mechanist, materialist and reductionist view of the world, and this has hindered and delayed the development of scientific studies about consciousness. In this wonderful book Mario Beauregard describes many studies about transcendent conscious experiences, nonlocal phenomena (beyond time and space) such as telepathy, remote viewing, precognition, presentiment, and mental influence on physical systems. He discusses how during meditation or in placebo treatment the mind can influence the function and structure of the brain (neuroplasticity), and how the mind can also change the activity of our genes (epigenetics). He writes about experiences of enhanced consciousness around the deathbed as well as after death, and also about proven cases of reincarnation. These studies support the idea of the continuity of consciousness after physical death. Postmaterialist science is based on the undeniable fact that consciousness is fundamental in the universe, and that it is a prerequisite for reality. We mostly do not realize anymore that the world, as we see it, only derives its "subjective" reality from our state of consciousness. Because it is only our consciousness that is determining how we perceive the reality in this world. So we have to change and expand our ideas about the nature of reality by changing our ideas about the function of the mind, with a better access to, and acceptance of, the "higher" dimensions of our consciousness. This will lead to a profound transformation of all aspects of life on our endangered planet. Highly recommended.

**Pim van Lommel**, cardiologist, NDE-researcher, author of *Consciousness Beyond Life*

# Expanding Reality

## The Emergence of Postmaterialist Science

# Expanding Reality

## The Emergence of Postmaterialist Science

Mario Beauregard

IFF
BOOKS

Winchester, UK
Washington, USA

JOHN HUNT PUBLISHING

First published by iff Books, 2021
iff Books is an imprint of John Hunt Publishing Ltd., No. 3 East Street, Alresford,
Hampshire SO24 9EE, UK
office@jhpbooks.com
www.johnhuntpublishing.com
www.iff-books.com

For distributor details and how to order please visit the 'Ordering' section on our website.

Text copyright: Mario Beauregard 2020

ISBN: 978 1 78904 725 7
978 1 78904 726 4 (ebook)
Library of Congress Control Number: 2020947595

A CIP catalogue record for this book is available from the British Library.

Design: Stuart Davies

UK: Printed and bound by CPI Group (UK) Ltd, Croydon, CR0 4YY
Printed in North America by CPI GPS partners

We operate a distinctive and ethical publishing philosophy in
all areas of our business, from our global network of authors to
production and worldwide distribution.

# Contents

# From the same author

*Brain Wars*, HarperCollins, 2012
*The Spiritual Brain* (with Denyse O'Leary), HarperCollins, 2007

# Foreword

## Gary E. Schwartz, PhD[1]

*If at first an idea is not absurd, then there is no hope for it.*
Albert Einstein

*May this book transform your life by finally liberating you from the materialist straitjacket.*
Mario Beauregard

The history of science reminds us that science is, first and foremost, a careful step by step process of creative discovery and application. As Ted Arroway, a character in the 1997 movie *Contact* said to his budding young daughter and future astrophysicist, Dr. Ellie Arroway, "small moves, Ellie, small moves."

However, every now and again, scientists take "small steps" that prove to be "giant leaps" for the evolution of human understanding and life in general. Thomas Kuhn, the distinguished historian and philosopher of science, called these revolutionary transformations in human understanding "paradigm changes."

Interestingly, in each instance the revolutionary "small" step/idea was "simple" and frameable in a single sentence even though its giant implications and applications were complex and far reaching. These revolutionary ideas and implications challenged the prevailing established beliefs, and they were typically perceived by most of the scientists and public in their day as being wrong, impossible, and even "absurd."

Historical examples of small/simple steps that have become giant/transformative leaps for humanity include:

1. The idea that the earth is round
2. The idea that the earth revolves around the sun
3. The idea that time and space are relative, and
4. The idea that seemingly solid matter is composed of quanta of energy whose properties defy common sense and experience

As Dr. Mario Beauregard clearly and convincingly presents in *Expanding Reality: The Emergence of Postmaterialist Science*, science is experiencing another paradigm change whose revolutionary implications may be even more transformative for science and society than 1-4 above. Not only are the implications of postmaterialist science far reaching for humanity, they have the potential to resolve the historic as well as contemporary war between science and religion, and in the process, enhance the prospect for global healing and peace.

The small step that is a giant leap for humankind is:

5. The idea that consciousness is a core property of universe, not only on a par with physical fundamental properties – e.g., information, energy, and matter – but even more primary.

It is virtually impossible to read Beauregard's comprehensive and engaging presentation of the emergence of postmaterial science and not see reality in a new, expanded way. Like Beauregard's previous books, *The Spiritual Brain* and *Brain Wars*, *Expanding Reality* provides a delicious four course meal of historical context, scientific evidence, personal perspective, and big picture analysis that leaves the reader in a state of wonder, awe, and hope.

I was inspired to contact Dr. Beauregard after reading his previous books. We quickly discovered how similar and complementary our emerging perspectives were, and I invited him to join me and Lisa Miller in organizing the International Summit on Postmaterialist Science, Spirituality, and Society held in Tucson in February 2014. This visionary meeting – Mario calls it a "Pivotal Event"– resulted in many practical collaborative products, including a book edited by Mario Beauregard, myself, Natalie Dyer, and Marjorie Woollacott titled *Expanding Sciences: Visions of a Postmaterialist Paradigm*.

It takes knowledge, skill, vision, inspiration, and courage to write a big picture book whose topics range from "Thoughts that Transform the Body" (Chapter 1) to "Journeys to the Source" (Chapter 8). Like other paradigm-changing books in the history of science, *Expanding Reality* will receive both praise and distain. Speaking bluntly, some readers will love this book and others will despise it. This extreme diversity of opinion comes with the territory, as the author of *Brain Wars* knows firsthand.

In my chapter in *Expanding Science* titled "Extraordinary Claims Require Extraordinary Evidence: The Case for Postmaterial Consciousness," I discuss what I call the Kepler Challenge as expressed by Dr. Carl Sagan, the distinguished astronomer and skeptic.

*When he [Johannes Kepler] found that his long-cherished beliefs did not agree with the most precise observations, he accepted the uncomfortable facts.*

*He preferred the hard truth to his dearest delusions, that is the heart of science.*

Dr. Beauregard encourages us to be genuine truth seekers. He invites us to practice – and celebrate – the "heart of science," and thereby be liberated from the straitjacket of materialism. Once again, it is time for us to expand our hearts and minds.

# Introduction: The Downfall of Scientific Materialism

*Consciousness cannot be accounted for in physical terms, for consciousness is absolutely fundamental. It cannot be accounted for by anything else.*
Erwin Schrödinger

## The Essence and Origin of My Research

Certain members of the general public, as well as journalists, sometimes inquire as to why I conduct research on consciousness, transcendental experiences and brain-mind relationships.[2] Until now, eager to protect my reputation, I have remained vague on the subject. As this no longer matters, I have decided to be honest and address those questions in this book.

As a scientist, empirical evidence related to my research has been crucial in the key thematic areas addressing specific sectors of my research program. This observation, or experimentation-based evidence, is analyzed in the chapters that follow. However, to be honest, my work as a researcher has been primarily influenced by significant first-hand spiritual experiences at various instances in my life.

Since I do not consider myself a guru and do not wish to risk marginalization, this book will not focus on these experiences. I do need, however, to mention some of them briefly to shed some light on me and my roots. The first one occurred when I was only eight years old.

My parents' farm was found near a vibrant and mysterious forest, which I explored every now and then. On one beautiful summer day back in 1970, I ventured into the entrancing woods. It was hot, and after walking for some time, I realized I was exhausted. I sat down on a large grey rock and gazed at the magnificent trees around me. After a few minutes, I started to feel a strong connection with the trees and the rock – I could sense that they were imbued with life. It felt like the rock, the trees and I were part of a much larger whole, much bigger than my "small self." My life's goal became very clear following this powerful experience – I would become a scientist who would help demonstrate that the human essence is not determined by the brain.

Some twelve years later, a new life cycle began for me. This new cycle was also marked by various determining spiritual experiences. It was new, as until then, I had been blessed with perfect health. Not anymore.

One January morning in 1982, I woke up to find that my body was not functioning like it did the day before, and that my visual perception was dramatically altered. I felt completely drained of energy, and my stomach, back, and multiple joints were hurting. I also felt dizzy and nauseated, and my breathing was labored. I was experiencing a kind of mental fog and my visual perception of the outside world was not the one I was used to. In fact, all the objects in my line of vision seemed to be continuously whirling.

I felt way too sick and weird to attend my classes at university. I did not understand what was happening, but I was convinced that it was something serious. The next evening, I called my parents to tell them about my predicament. They asked me to return home as quickly as possible – which I managed to do the very next day, after gathering all my willpower, but not without difficulty and misery.

I stayed in bed for almost a year. Weak and barely feeding myself, I had no other choice than to temporarily drop out of university.

My parents were desperate and utterly helpless. We agreed that we had to find the underlying source of my condition, which might be the solution to my problem.

My mother took me to see several medical specialists,

including a neurologist, a psychiatrist, an ophthalmologist, a gastroenterologist and an internist. Some of these doctors decided not to further the investigation, dismissing me as a hypochondriac. One doctor, claiming I was showing signs of schizophrenia, prescribed an antipsychotic which I threw out right away. The other medical specialists, puzzled by the mysterious accumulation of my symptoms, subjected me to a battery of tests – all negative.

I could feel myself slowly withering away and I could not reconcile what I was going through with the experience that had left such a mark on my childhood. It did not seem to make any sense and it filled me with bitterness – I would never be able to accomplish my life's mission in such a deplorable state. I found myself in a very dark place and suicidal thoughts started to creep into my consciousness – I could not live like this anymore.

One evening, I was so desperate that I mentally bellowed at the sky. Apparently, my supplication did not fall on deaf ears, as a few days later, in the middle of the night, I suddenly felt that I was leaving my physical body through my heart. I then sensed a Being of Light, radiating immense and unconditional love. The Being of Light reassured me telepathically that what I was experiencing was no disease, but rather a process of transmutation. He also told me that I was not alone and that I had to hang in there. Moreover, to restore my confidence, this Being of Light told me about events that would take place shortly, and he also mentioned that the severity of my symptoms would gradually decrease over the following months. All the Being of Light's predictions came true. Many other things happened during this life-defining episode, but I will not mention them as the main purpose of this book is not to share my personal experiences.

After that, I mustered enough strength to resume my studies, and my symptoms faded slowly but surely, just like the Being of

Light had predicted. Still, getting back on my feet was no easy task.

Seven years after my health problems first started, a friend introduced me to a famous doctor and microbiologist at the Hôtel-Dieu Hospital in Montreal. The latter performed a thorough battery of medical tests. He detected the presence of viral agents in my body – the Epstein-Barr virus, cytomegalovirus and Coxsackie virus. Together, these viruses could explain most of my symptoms. I also learned that I had cerebral vasculitis, an inflammation of the blood vessels of the brain. The severity of the viral infections left the microbiologist bemused as to how I had managed to remain on my feet and continue with my studies.

It took me twelve long years to get back in perfect health. During this life cycle, my psychic faculties developed remarkably, and I have remained in close contact with the spiritual world since this lengthy episode.

I saw the microbiologist again one last time when everything was over. He told me that there was no explanation for my remission, which was simply "miraculous," from a medical point of view.

## Science is Not Tantamount to Materialism

It may seem surprising to many readers, but most contemporary men and women of science are unaware that the "modern scientific worldview" is largely based on metaphysical assumptions, that is, assumptions about the nature of reality.[3] These assumptions were first put forward by ancient Greek philosophers who preceded Socrates. These philosophers, referred to as "Pre-Socratic," proposed several interesting ideas and theories that are now available to us.

At first glance, this crass ignorance, which also characterized many scientists from previous generations, may not seem so important. However, this is not the case, as such ignorance has had considerable negative impacts. Indeed, this ignorance is largely responsible for the delay of science in the fields of research related to human interiority. This becomes evident if we take as a point of comparison scientific fields geared towards technological development. Furthermore, it is this ignorance that has made most scientists prisoners of an ideological trance that has lasted a few centuries. This is hardly surprising since knowledge, especially when disinterested, is the best antidote against ideology.

Several centuries after the Pre-Socratics, the philosophical postulates, which formed the vision of the modern scientific world, became associated with classical physics. These postulates include *materialism,* the idea that matter is the only reality and that everything in the universe is composed of sets of material particles and physical fields[4] – as well as *reductionism,* the notion that complex things can only be understood by reducing them to the interaction of their parts or to simpler or more basic things, such as material particles. Other postulates include *determinism* – the idea that future states of physical or biological systems can be predicted based on their present states – and *mechanism,* the notion that the world functions as a huge machine determined by immutable physical laws.

For the founding fathers of classical physics – such as Galileo, Descartes and Newton – these philosophical postulates were useful hypotheses that could guide them in their exploration of the material world. However, these pioneers of modern science, who were also spiritual men, did not believe that the world could be reduced only to its material dimension. But this crucial nuance ends up being forgotten by their successors, and during the nineteenth century, the postulates in question were changed into dogmas and united to form a system of beliefs that became known as "scientific materialism."

This belief system implies that our consciousness and everything we experience subjectively – for example, our perceptions, our thoughts, our emotions, our memories, our free will, our sense of personal identity and our spiritual illuminations – is identical or can be reduced to electrical and chemical processes taking place in our brain. Moreover, scientific materialism posits that our thoughts cannot affect our brain, body, actions, and the physical world. In other words, human beings are complex biophysical machines only, and therefore, our personality and our consciousness return to oblivion when we die.

Materialist ideology became extremely dominant in the academic milieu during the twentieth century. So dominant, in fact, that most scientists started to believe that this ideology was based on empirical evidence and that it represented the only possible rational conception of the world. Scientists also tacitly understood that they could endanger their careers if they dared to question this dogmatic and intolerant ideology.

We must admit that scientific methods based on materialist philosophy have proved highly successful as they have led to a better understanding of nature, as well as greater control and freedom through technological advances. However, the almost absolute dominance of materialism in the academic world has greatly stifled science and hindered the development of the scientific study of the mind, consciousness and spirituality. Faith in this ideology, as an exclusive explanatory framework of reality, has led scientists to neglect the subjective dimension of human experience. Scientific materialism has also brought a severely distorted and impoverished conception of ourselves and of our place in nature.

Science is first and foremost a non-dogmatic and open method of acquiring knowledge about nature. This method is based on observation, experimental investigation and theoretical explanation of phenomena. In this sense, science is not tantamount to materialism and should not be influenced by any belief, dogma or ideology.

## The "Dematerialization" of the World

In April 1900, British physicist Sir William Thomson, also known as Lord Kelvin, asserted with confidence, at a conference he gave at the Royal Society of London, that there was practically nothing left to be discovered in physics. Lord Kelvin, however, recognized that two "little clouds" darkened the serene sky of his discipline – the problem of ether and that of blackbody radiation. He also claimed that these minor problems, which were anomalies for classical physics, would soon be solved by refining the existing theories. Lord Kelvin could not be more mistaken as these two little clouds shook physics to its very core – the ether problem was eventually solved by the theory of relativity while that of blackbody radiation was solved by Max Planck who proposed a bold hypothesis that soon led to the birth of quantum mechanics (QM), a revolutionary branch of physics.

This new physics invalidated the metaphysical postulates underlying scientific materialism by demonstrating that atoms and subatomic particles are not solid objects – they do not exist in definite places and times. Indeed, in the quantum domain, atoms and subatomic particles show "tendencies to exist,"[5] shaping a world of possibilities rather than things or facts. QM has thus "dematerialized" the world by showing that it is not at all composed of small grains of matter comparable to tiny billiard balls.

QM also demonstrated the need to introduce the mind into its basic conceptual structure since it has been discovered that the particles observed and the observer – the physicist and the method used for observation – are related. This is called "the effect of the observer." This phenomenon reveals that the physical world is not the only or primary component of reality, and it cannot be fully understood without referring to the mind. In other words, as Wolfgang Pauli, one of the founders of QM, said, our reality consists of two complementary but distinct aspects: the physical and the psychological, *physis* and *psyche*.

The effect of the observer has led some of the pioneers of QM – for instance, Max Planck, Erwin Schrödinger, John von Neumann and Eugene Wigner – to propose that the consciousness of the observer is vital for the existence of measured physical events, and that mental events can influence the physical world. The results of recent studies support this interpretation (we will come back to this topic later).

Non-locality (or non-separability) is another important discovery in QM. This principle is based on entanglement, which refers to persistent instantaneous connections between particles (e.g., photons or electrons) that interacted together before being separated. Surprisingly, these connections persist even if the particles in question are separated by huge distances (for example, billions of light-years). This counterintuitive aspect of nature, which was described by Albert Einstein as a "spooky action at a distance," has been confirmed experimentally in several laboratories since the 1970s. Non-locality and entanglement suggest that the universe is an indivisible whole. As I will demonstrate in this book, the principle of non-locality does not seem to be confined to the microphysical domain since it also exists in the world of the *psyche*.

It is a pity that even though QM has invalidated the metaphysical postulates of materialism, some scientists and philosophers, by ignorance or bad faith, still cling to this superstition. Some of them, the "fundamaterialists,"[6] behave like true religious fanatics to defend this erroneous and obsolete system of beliefs.

## A Pivotal Event

A few years ago, in collaboration with psychology researchers Gary Schwartz from the University of Arizona and Lisa Miller from Columbia University, I helped organize the International Summit on Postmaterialist Science, Spirituality and Society.

The purpose of this meeting, held in Tucson in February 2014, was mainly to analyze the various types of empirical evidence supporting postmaterialist science. There was also a discussion about the evolution and recognition of a postmaterialist paradigm, as well as its potential impact on spirituality and society. Internationally-recognized scientists, from various fields of expertise – including biology, neuroscience, psychology, medicine and parapsychology – took part in this pivotal event.

The conclusions of the summit were published a few months later in the form of a document entitled *Manifesto for a Post-Materialist Science*. This manifesto was published in the scientific journal *Explore: The Journal of Science and Healing*. Since its publication, more than 300 scientists and philosophers from around the world, including some very prestigious ones, have signed this document.[7]

Some have criticized this manifesto by saying that my colleagues and I are non-conformist scientists, and that our point of view does not represent the dominant view. However, as noted in a highly complimentary article about us by Dave Pruett[8] – a former NASA researcher and Emeritus Professor of Mathematics at James Madison University: "We must not forget that neither Copernicus nor Galileo nor Kepler nor Einstein represented the current and dominant scientific position. These great minds questioned the scientific status quo, and we eventually ruled in their favor and accepted their theories."

Our postmaterialist science summit also led to the *Open Science Campaign* and the creation of a website of the same name. The site is a portal promoting research that goes beyond the dogmas that still significantly influence contemporary science. The main areas of research covered include studies on consciousness, integrative medicine and healing, postmaterialistic scientific approaches, as well as novel aspects of cosmology, physics, chemistry and biology. The website refers to books and provides access to videos, articles and links to sites of researchers and organizations known for their open-mindedness.

More recently, I participated in the writing and editing – in collaboration with Gary Schwartz, Natalie Dyer, a young neuroscientist at Harvard University, and Marjorie Woollacott, another neuroscientist – of an anthology entitled *Expanding Science: Visions of a Postmaterialist Paradigm.*[9] Several researchers, visionary and innovative, from various disciplines (e.g., Amit Goswami, Rupert Sheldrake, Larry Dossey, Pim van Lommel, Gary Schwartz and Dean Radin) as well as a few philosophers contributed to this work. Its main objective is to analyze the empirical evidence challenging the materialist position. This anthology discusses, from multiple stances, the emerging postmaterialist paradigm and its implications, and the broadening conceptual scientific framework. This important work targets nonspecialized readers.

A short time ago, I attended yet another important meeting, also held in Tucson, Arizona. This meeting marked the creation of *the Academy for the Advancement of Postmaterialist Sciences (AAPS).* One of the goals of this organization is to promote cultural change in connection with our understanding of human experience and reality. To achieve this, the academy encourages scientists, in various ways, to go beyond the limits imposed by an exclusively materialist and reductionist approach to science.[9]

Along the same lines, a few years ago, the Scientific and Medical Network (SMN, www.scimednet.org) – a worldwide professional community and membership organization for open-minded, rigorous and evidence-based enquiry into themes bridging science, spirituality and consciousness – created the Galileo Commission. The main goal of this project is to find ways to expand science so that it can explore and accommodate important questions and human experiences that mainstream science is unable to integrate. Following widespread consultation with 90 advisers representing 30 universities worldwide, the SMN published the Galileo Commission Report, a groundbreaking document (entitled *Beyond a Materialist Worldview: Towards an Expanded Science*) written by Professor Dr. Harald Walach, a researcher at the interface between medicine, psychology and consciousness studies.

## The Reasons behind This Book

Even though QM appeared almost a century ago, most people – and even scientists who are not physicists – unfortunately remain unfamiliar with its fascinating and profound implications. They are therefore unaware that this branch of physics demonstrates that the metaphysical postulates of scientific materialism are not valid. Therefore, a significant part of society remains under the influence of a mechanist, materialist and reductionist view of the world. For example, different spheres of human activity – like politics, economics, education, health and the media – are still guided in part by this outdated concept which we need to get rid of.

This is a propitious situation for some materialists who are also militant atheists, such as scientists Richard Dawkins, Lawrence Krauss, Richard C. Lewontin and Paul Zachary Myers, as well as philosophers Daniel Dennett, Patricia Churchland and André Comte-Sponville. These advocates of rationalism are actively involved in a cultural war against religion and claim to represent ultimate cognitive authority. They maintain the naive and presumptuous belief that neuroscience will eventually be able to completely reduce the mind and consciousness to what is happening in the brain. What should be noted about this is that this prophetic belief, which philosopher of science, Karl Popper, called *"promissory* materialism," had already been professed by proponents of atheistic materialism during the eighteenth century.

Materialist theories miserably fail to explain how the brain could produce mental functions and consciousness. These theories are also unable to explain what the philosopher David

Chalmers calls "the hard problem of consciousness": why and how the experiential aspects of our mental lives (which philosophers call the *qualia*), which we access by introspection, result from the activity of a group of neurons in the brain? Why do certain physical states of the brain cause lavender to look purple or for it to be painful to be stung by a wasp?

To better understand the mind-brain relationships, we must leave the suffocating and limiting framework of scientific materialism and consider all research on consciousness. These highlight a whole array of empirical evidence which, like discoveries in QM, are totally incompatible with the old materialist paradigm.

This empirical evidence is meticulously analyzed in this book. This includes the mental ability to influence brain activity, psychosomatic network and genes; psi phenomena such as telepathy, distance vision, precognition, mental influence on physical systems and living organisms; phenomena associated with death such as near-death experiences, shared death experiences, and deathbed visions; reincarnation; and transcendent experiences. In the last section of this book, I present my view of the main elements that make up the postmaterialist paradigm. Many of my colleagues and I are convinced that this paradigm will lead us to the next great scientific revolution. The implications of this new paradigm are therefore tremendous, both for everyday life on an individual level and for the future evolution of our species.

May this book transform your life by finally liberating you from the materialist straitjacket.

## Chapter 1

# Thoughts that Transform the Body

*Your worst enemy cannot harm you as much as your own thoughts, unguarded. But once mastered, no one can help you as much.*
The Buddha

### A Few Fundamental Dogmas in Neuroscience

Every human being – religious, scientist or atheist – needs to believe in something. In the case of atheists, they believe in "nothingness" after death, even though they hold no evidence of it. They also believe that God – or a transcendent principle at the origin of the world – does not exist. With respect to scientists and Science, and with science being a product of human activity, it is not immune to prejudices and dogmatic beliefs. This is particularly obvious when most scientists who work in a specific area of research embrace a dogma extensively. The impenetrable truth status attached to Darwinism in biology proves the impact that such a phenomenon can have.

For more than a century, neuroscience researchers have believed that new neurons could not develop in the adult human brain – we were born with a maximum number of neurons, and this would only dwindle throughout our life. Neuroscientists also believed that the adult brain was a static machine that did not have the ability to change. However, in recent decades, this dogma has been cast aside in light of several studies that have convincingly shown that parts of the human brain, as well as many animal species, retain the ability to produce new neurons throughout adult life. According to these studies, contrary to what we thought, the adult human brain is very plastic. Indeed, it constantly alters its structure and function by creating new neurons and synaptic connections between neurons. In addition, in the mature brain, existing neural networks are constantly reorganized, and new networks are developed. This neuroplasticity indicates that we are not prisoners of the brain we inherited at birth.

As mentioned in the introduction, another fundamental dogma in neuroscience, one that remains influential, is the idea that all mental events, consciousness and the self are simply reducible to the physical and biological processes of the brain. In this regard, different materialist theories have tried to explain how the brain produces the mind. One of them is the theory of psychophysical identity. This theory claims that we comprehend our mental processes and our consciousness in the first person, that is, from within and subjectively ("I feel happy."); while with neuroscience techniques, our brain activity is measured in the third person, that is externally and objectively ("My brain releases more serotonin when I am happy."). In other words, mental events and brain events are perfectly parallel, like two sides of the same coin. However, it is brain states, electrical impulses, and chemical reactions in our brain that create mental states, not the other way around. Most contemporary neuroscientists stand by this view, including Gerald Edelman and Jean-Pierre Changeux.

Eliminativism is another prominent materialist theory about the relationship between mind and brain. The philosophers Paul and Patricia Churchland, as well as Daniel Dennett, are illustrious supporters of this radical theory that denies the existence of mental events: our mental world is just an illusion, and we are only imagining having sensations, memories, emotions and thoughts. Eliminativism recognizes only the biophysical processes of the brain. It posits that "mind," "consciousness," "ego," and "free will" are prescientific concepts that stem from naive and simplistic ideas belonging to "popular psychology." Proponents of this theory hope that these concepts will soon be eradicated thanks to advances in neuroscience. As to *qualia* – how things appear to us individually, Daniel Dennett vehemently claims that these experiential aspects of our mental lives are also mere illusions.

Those who support materialist theories often incorrectly claim that methods in neuroscience – like, for example, brain imaging techniques – measure mental events. This cannot be more wrong because these events are not physical. Indeed, a thought has no shape, no mass, and no volume. Materialists also often claim that studies using these methods demonstrate that the brain generates both mind and consciousness. They believe that the fact that mental functions are affected when the brain is damaged strongly supports this concept.

This conclusion is inaccurate because neuroscience methods only allow us to measure correlations – that is, reciprocal relations – between cerebral states and mental events. Hence, when the state of the brain changes, mental activity also changes; and when mental activity changes, the state of the brain also changes. However, the correlations between mental activity and brain activity do not imply causality and identity. For example, researchers can record the electrical activity of the brain using electroencephalography (or EEG). Electrodes are applied to the scalp of subjects while they experience a feeling induced by emotionally charged images. However, the brainwaves recorded by the EEG are completely different from the emotions that are experienced by the subjects. Moreover, the correlation between the electrical activity of the brain and the variation of the emotional state of the subject does not mean that the changes in the EEG are the cause of the changes reported in the emotional state.

The flawed belief about the identity of mental events and their neural correlates leads to what has been called the "mereological fallacy": the erroneous attribution of mental properties to parts of the brain or to the brain itself.[10] The materialists who commit this error elevate the brain to the level of an omnipotent physical entity, whose properties make it possible to explain all mental phenomena. These materialists do not seem to realize that it is the person – not their brain – who is conscious, who thinks, who decides and who believes.

## The Compelling Placebo Effect

David Kallmes is an affiliated neuroradiologist at the renowned Mayo Clinic, whose headquarters are in Rochester, Minnesota. During the last two decades, Kallmes has performed many vertebroplasties. Conducted under continuous radiographic control, this surgical technique allows for the restoration of a vertebral fracture by injecting a type of bone "cement." The cement is completely hard after a few hours, and twelve hours later, the patient can stand up.

Vertebroplasty produces impressive results even when the wrong vertebra is accidentally filled with cement. This peculiarity intrigued David Kallmes. Suspecting that other factors were responsible for the effectiveness of this medical procedure, he decided to conduct a study to determine if vertebroplasty was more effective than a placebo – a treatment that has no biological action but can be effective when the person receiving the placebo thinks they are receiving active treatment.[11]

Doctors are aware that the beliefs and expectations of their patients, when it comes to treatments, may affect the results of these treatments. In clinical trials, medical researchers consider the placebo effect a nuisance since this phenomenon prevents the conclusion of whether the new experimental treatments are effective. To circumvent this problem, participants in clinical trials are randomly assigned to either the group receiving the true (active) treatment, or to the one receiving the fake treatment (placebo, for example, a sugar pill). Neither researchers nor participants have access to this information before the end of the tests. This is called the double-blind randomized clinical trial protocol, with a control for the placebo effect. When the clinical trial is complete, researchers subtract the placebo effect from the real treatment.

In this particular case, Dr. Kallmes and his colleagues developed a clinical trial in which some patients would receive true vertebroplasty while other patients would receive placebo surgery. To ensure that the 130 participating patients at the clinical trial could not know if they received the active treatment or the false treatment, they were prepared for their "operation" in the same way: they were brought into the operating room and then they were injected with an anesthetic agent in the back. Then a computer program randomly decided which patients would receive vertebroplasty or placebo. For both types of procedure, the doctors opened the container containing the bone cement, which gives off a strong smell akin to nail varnish solvent.

Half of the patients received vertebroplasty, while the other half received the fake surgery. For the placebo patients, the doctors adhered to the following scenario: they pressed the patients' backs and told them: "The cement is entering now, everything is fine, a few more minutes and everything will be over."[11]

Bonnie Anderson was one of the patients recruited for this clinical trial. She had a broken vertebra after a fall in her kitchen. Bonnie could hardly move due to the acute pain, and she could only walk by holding on to things. However, one week after the surgery, this 76-year-old woman could play golf again. What is remarkable in her case is that Bonnie received the fake surgery. The placebo procedure was also effective for several other patients who participated in this clinical trial. So effective that there was no statistically significant difference in terms of decreased pain relief and functional improvement between patients in both groups.

The research results indicate that the placebo effect is

involved in all types of medical and psychological treatments. Thus, the meta-analyses – which are statistical analyses combining the results of several studies to more accurately assess the true magnitude of the phenomenon being studied (or "effect size") – indicate that the placebo effect plays a crucial role in clinical trials for drugs targeting mood disorders such as major depression. In this regard, clinical researchers and psychologists Irving Kirsch and Guy Sapirstein examined the results of more than 19 clinical trials involving more than 2,000 patients. These researchers found that 75% of therapeutic outcomes are attributable to the placebo effect.[12]

Research on the placebo effect also reveals that the more we believe in a given treatment and trust a therapist – it does not matter if it's a Western medicine doctor, a homeopath or a shaman – the more likely it is that the treatment will be effective and mobilize our powerful innate mechanisms of self-healing. Hope, positive emotions, motivation, anticipation of improvement, and a warm and thoughtful attitude on the part of the therapist also contribute to induce a strong placebo response.

Administering a placebo does not always trigger the desired self-healing mechanisms. It can sometimes lead to unwanted and unpleasant symptoms such as drowsiness, nausea, fatigue and insomnia. This phenomenon is called the nocebo effect and represents the dark side of the placebo effect. It can occur when our expectations of a treatment are negative rather than positive. If the administration of a treatment is accompanied by warnings related to potential side effects, these negative effects are more likely to occur. In the book *Love, Medicine and Miracles*, Dr. Bernie Siegel, a now retired pediatric surgeon, cites a study that aimed at testing a new chemotherapy drug. Patients randomly assigned to the placebo group were informed that they could receive this new drug, and that it could have negative side effects. Even though these patients received only one saline injection, 30% of them lost their hair.[13]

The nocebo effect is also observed with real drugs. For example, in a clinical trial, finasteride was given to men suffering from prostate enlargement. Half of the participants in this clinical trial were warned that this drug could produce erectile dysfunction, while the other half of the participants were not informed of this potential side effect. In the informed group, 44% of participants reported that they were experiencing erectile dysfunction, while this side effect was reported by only 15% of participants in the uninformed group.[14]

The nocebo response is not unique to clinical trials. Thus, after the sarin gas attack on the Tokyo subway in 1995, hospitals in this Japanese megalopolis were overwhelmed by cases of individuals who believed they had been exposed to this deadly neurotoxic substance. These individuals had symptoms that had been highly publicized, such as nausea and dizziness. It turned out that these individuals had not really been exposed to sarin. This type of reaction is common when the agent is not visible, such as radiation and some chemicals.[15]

The different types of placebo reactions are associated with various changes in brain activity. This phenomenon is well illustrated by the cerebral response of patients suffering from Parkinson's disease, a degenerative neurological disorder characterized by tremors, slowing of mobility and muscle rigidity. In this neurological disease, there is a marked decline in the amount of dopamine, a chemical messenger, found in basal ganglia (also called basal nuclei). In a positron emission tomography (PET) scanner, researchers scanned the brains of Parkinson's patients to compare brain responses to apomorphine – a drug that activates dopamine – and salt water – a placebo treatment. Both treatments were administered following a double-blind protocol. Placebo treatment was presented as stimulating mobility. Compared to those who received no treatment, placebo patients showed a dramatic increase in dopaminergic activity in the basal ganglia. This increase was comparable to that observed with the therapeutic dose of apomorphine given to other patients.[16]

The nocebo response is also accompanied by significant changes in brain activity. For obvious ethical reasons, very few neuroimaging studies have been conducted to examine the brain's response to negative expectations. The few studies that have been conducted in this regard focus on pain. These studies reveal that negative expectations lead to an amplification of pain – the more the degree of anticipated pain increases, the more the brain regions involved in the perception of pain become activated. Conversely, expectations of decreased pain reduce the activation of brain regions related to the treatment of pain information.[17]

## Managing Emotions

Emotions are physiological changes in terms of heart rate, blood pressure, muscle tension, skin temperature, etc. in response to events occurring either in the external environment (for example, seeing a wild animal close to us in a split second while strolling in the forest), or in the internal environment, that of the *psyche* (for example, the recollection of a memory that is dear to us). These multidimensional responses also involve a cognitive analysis of what is happening in the environment (internal or external), behavioral expressions (for example, defending oneself in response to a threat) as well as changes in subjective experience – that aspect of emotions called feeling (for example, being scared).

Our emotions play a crucial role in our life – they help us make choices about the people, events and situations we encounter during our lifetime. However, when our emotions are repeatedly negative, multiple problems and enormous suffering may ensue.

The physiological aspects of emotions are closely related

to brain regions that are part of what is commonly called the emotional brain (in neuroscience jargon: the limbic system). These areas are ancient, from a phylogenetic point of view, and are found in all mammals.

Luckily, most members of our species can learn to control their emotions. Otherwise, the innermost impulses in our archaic emotional brain would completely overwhelm us. This vital capacity for the future of humanity is mediated by some subdivisions belonging to the prefrontal lobe, the most recent brain structure in terms of our biological evolution. Compared to other mammals, the prefrontal lobe is much more developed and bulkier within our species. This structure, sometimes called the "Organ of civilization," is considered as the key region for implementing the rational aspect of the mind.

There are important anatomical connections between the various regions of the prefrontal lobe and those that are part of the emotional brain. Connections from the prefrontal lobe to the emotional brain transmit inhibitory information, which allows voluntary control of emotions and feelings that accompany the activity of limbic structures. This capacity is one of the main components of emotional intelligence.[18]

At the beginning of the new millennium, I was the first neuroscientist to undertake a research program aimed at studying what goes on in the brain when healthy individuals try to voluntarily take control of their emotions. In the initial study of this research program, my colleagues and I asked young men to watch clips from erotic films. Participants were scanned with functional magnetic resonance imaging (fMRI) – a method to measure changes in blood oxygenation connected to neuronal activity – during a control condition and an experimental condition. In the control condition, participants only watched movie clips and had to react normally. In the experimental condition, the participants had to cognitively distance themselves – they had to watch other extracts of similar films in a dispassionate way, without trying to evaluate or judge them. A questionnaire was used to measure the intensity of the emotional responses in both conditions. This questionnaire revealed that all the participants had been sexually excited by the clips of erotic films. Unsurprisingly, in the control condition, sexual arousal was correlated with the activation of various brain structures that are part of the emotional brain (for example, the amygdala and the hypothalamus). What is remarkable in this study is that even if the participants had been trained in cognitive distancing only for a few minutes before the start of the experiment, they were all able to subjectively reduce their excitation during the experimental condition. Moreover, no activation was detected in regions belonging to the emotional brain. As per the notion that the prefrontal lobe acts as a regulating center of emotions, an activation was recorded in some subdivisions of this brain structure.[18]

In the second stage of our research program, we conducted another fMRI study using a similar approach.[19] In this second study, we measured the brain activity of young women trying to master a temporary response of sadness, which was also evoked by film clips. These showed the death of a loved one. The results were similar to those of the first study. The activity decreased significantly in the emotional brain during the experimental condition, while there was increased activity in some areas of the prefrontal lobe.

These results show that an adult in good mental health is not just a simple biological machine that only reacts to various stimulations. In fact, this person is neurobiologically equipped to be able to control her reactions and those of her brain in the face of emotionally charged events. It is important to point out here that even if the neurobiological equipment in question is in the prefrontal lobe, this crucial structure of the brain must be directed consciously and voluntarily by the person.

As part of this research program, I sought to determine

whether mentally and neurobiologically healthy individuals are also able to influence the activity of chemical messengers that play a significant role in emotional life. My colleagues and I used PET to evaluate serotonin production during rapid and sustained changes in an emotional state.[20] This chemical messenger is known for its central role in controlling emotions and mood fluctuations.

We asked participants in this study to experience temporary states of joy and sadness through self-induction. We suggested that they relive intense emotions associated with specific autobiographical memories. All participants indicated that they had experienced both emotions. The PET results reflected these subjective perceptions since the level of sadness felt was correlated with a reduction in serotonin production in the emotional regions of the brain; while conversely, the level of joy felt was correlated with an increase in serotonin production in areas of the emotional brain. These results are consistent with studies demonstrating a reduction in serotonergic activity in these areas in individuals with major depression.

## Brain Transformation through Meditation

Meditation refers to a wide range of mental training techniques that have been developed to promote cognitive abilities, emotional balance and well-being. During the last decade, several studies have indicated that intentional directing of attention through meditative practices can lead to significant plastic changes in the brain. These practices are generally classified into two categories: concentration and mindfulness.[21]

Concentration practices involve focused attention on certain specific body sensations (breathing, for instance) and aspects of mental activity (for example, a mental visual image or a repeated sound).

As for mindfulness practices, they entail letting all thoughts, emotions and sensations emerge from moment to moment, while maintaining, without judging nor analyzing, an attentive yet detached consciousness. These two meditative technique categories slow the internal mind monologue down – a deeply soothing feeling of serenity as well as an expanded self-experience which is no longer centered on the thoughts of the meditator and representations of her body.

The "mindfulness-based stress reduction" method (MBSR) is an eight-week group program based on the practice of mindfulness meditation. Thousands of people have participated in this program, and studies have shown its effectiveness in reducing stress, anxiety and depression, and increasing well-being, compassion and spirituality.[22]

Research with the EEG reveals an increase in alpha and theta activity during mindfulness meditation practices, such as Zen and Vipassana. Alpha waves (8 to 12 Hz [cycles per second]) are generally associated with relaxation, whereas theta waves (from 4 to 8 Hz) are more specifically associated with meditative states. It seems that the extent of the increase of theta waves during meditative states reflects the level of expertise of the meditators.[23] Other EEG studies indicate that the increase in beta 2 activity (from 20 to 30 Hz) and gamma activity (from 40 to 100 Hz) is a characteristic of concentrative meditation states. Beta 2 activity is often observed in tasks involving focus of attention. Gamma activity is believed to be related to the content of our mental experiences and to our consciousness.[24] Gamma waves were recorded in experienced Buddhist Tibetan monks while they were engaged in a form of meditation called "non-referential compassion."[25] In this type of meditative state, meditators focus on boundless compassion and loving kindness for all living beings. Stunning increases in gamma waves were measured during the non-referential state of compassion in relation to a resting state. The more the monks have trained in this form of meditation, the higher and longer their gamma activity is. On the other hand, this high gamma activity lingers among monks even though they have stopped meditating. These results suggest that even outside a meditative state, meditation training can have long-term changes on brain activity.[26]

All EEG meditation studies carried out so far support the idea that different neuroelectric signatures are associated with various types of meditative practices. This is not surprising considering that these practices differ in their subjective experiences and mental processes.

FMRI has also been used in several studies to identify neuroplastic changes induced by the practice of meditative techniques. Some of these studies were conducted by my research team. In one of them, we examined the effects of mindfulness meditation on brain responses to emotional images.[27] In this study, we also analyzed the impact of the duration of the training in mindfulness meditation on emotional brain responses to such images. Experienced meditators, followers of Zen meditation, were compared with novice meditators. These novice meditators practiced mindfulness meditation for 20 minutes a day during the week before the experiment. Both groups of participants were scanned while viewing the images in a state of mindfulness or in a state without mindfulness. From the experiential point of view, the participants in both groups perceived the images seen in a state of mindfulness as less intense than those seen in a state without mindfulness. Moreover, in the experienced meditators, mindfulness was accompanied by reduced activity in emotional brain areas. These results suggest that training in mindfulness meditation leads to a reduction in brain activity, accompanied by a calming down of the mental state.

Other fMRI studies have shown that the practice of meditation leads to changes in the functioning of the brain regions involved in attention and concentration, as well as empathy.

It is now well established that the structure of an adult brain can change following repeated practice. For example, there is an increase in the thickness of the areas associated with music in the brain of musicians, and visual and motor areas in the brain of jugglers. In this context, some researchers have wondered whether the practice of meditation can also lead to structural changes in the brain. These researchers used a technique called anatomical MRI to tackle this question. The studies revealed that the training in mindfulness meditation leads to an increase in the density of grey matter – which contains the cell bodies of neurons – in areas of the brain involved in learning and memory, empathy and emotion management.[28] The practice of meditation can also change the white matter, which is responsible for communication between different regions of the brain, in brain areas which are important for attention and the management of emotions.[29]

## Impact on our Psychosomatic Network

Immunology researchers have long argued that the immune system and the nervous system are separate. However, about 30 years ago, research indicated that the immune system is not really an "isolated" defense system because chemical messengers produced by immune cells inform the brain, and the brain also sends messages to the immune system. Since then, it has also been discovered that the nervous system and the immune system interact with the endocrine system. These discoveries led to the emergence of psychoneuroimmunology (PNI), the study of interactions between mental processes and these three physiological systems. All these together make up the "psychosomatic network," and communicate with each other bi-directionally, through a common biochemical language that involves the sharing of receptors as well as neurotransmitters, cytokines (molecules produced by the cells of the immune system) and hormones. These act as "informational substances."[30] It is *via* this vast multidirectional network that our thoughts, emotions and subjective experiences are chemically translated, and that they influence our health and well-being. PNI studies indicate that it is through the interaction between psychological factors and biochemical changes affecting the psychosomatic network that the causes, development, and outcomes of a disease are determined.

Let's consider stress for a moment. When we stress, our adrenal glands secrete cortisol, and our brain releases chemical messengers such as adrenaline and norepinephrine. A little stress can help us mobilize all our resources to achieve our goals. However, too much stress can perniciously affect immunity. For example, acute or chronic stress can increase the duration and severity of infectious diseases, reactivate latent viruses, and prolong wound healing. This is well illustrated by the fact that when medical students receive a hepatitis B vaccine, during a very stressful final exam period, they do not develop complete protection against this life-threatening viral infection.[31] Situations and events that are perceived as uncontrollable, such as unemployment and natural disasters, can also weaken immune responses.[32] In addition, chronic negative emotional states, such as those noted in major depression, are associated with increased mortality.[33] Increased risk of death and decreased immunity are also observed in bereaved spouses during the first years after the death of their partner.[34]

Unlike negative emotions, positive emotional states can significantly improve health. In another PNI study, researchers administered the influenza virus to participants who were then placed in quarantine, and the reactions to the virus were analyzed according to the emotional state of the participants. The results showed that participants whose emotional state was positive had a reduced risk of developing influenza.[35] A positive emotional state is also associated with a decrease in mortality.[36]

Most of the time, we influence our psychosomatic network and our health unconsciously and involuntarily. However, some research suggests that it is also possible to have a conscious and deliberate influence on the activity of our psychosomatic network. In this regard, several studies indicate that in cancer patients, relaxation and mental imagery – the ability to form mental images of objects or events that are not in the sensory field – reduces anxiety and depression, reduces nausea and vomiting caused by chemotherapy, and increases the activity of the immune system and the quality of life.[37]

David Seidler's story reveals how mental imagery can sometimes be effective in fighting serious diseases. This Anglo-American playwright and screenwriter is best known for *The King's Speech* screenplay, for which he won the 2011 Academy Award for Best Original Screenplay. In 2005, at almost 70, Seidler learned he had bladder cancer. After seeing his urologist, Dr. Dino DeConcini, Seidler decided against chemotherapy or having his bladder removed. Instead, he took supplements to boost his immune system. He also opted for the surgical removal of the cancerous tumors in his bladder.

A few months later, David Seidler learned that his cancer was back. He made an appointment with Dr. DeConcini to have the cancerous tumors removed once more. This appointment was set for a few weeks later. His wife at the time suggested that he try to visualize his cancer disappearing. At first, Seidler thought this suggestion was the most ridiculous thing he had ever heard. However, after further contemplation, Seidler decided he had nothing to lose.

For two weeks, he visualized a bladder in perfect condition for hours. Then he went to meet Dr. DeConcini for his second surgery. The urologist was astounded to find that Seidler's cancer had completely vanished. This cancer has never come back since. David Seidler is convinced that he survived his bladder cancer because he did not give in to self-pity, and instead, used the same vivid imagination he did to write movie scripts.[38]

## We are Epigenetic Engineers

There is growing evidence that the mind can also influence genes. With this in mind, I think it would now be relevant to include genes in our representation of the psychosomatic network.

During the twentieth century, several influential biologists have defended genetic determinism – the idea that physiological processes of the body – as well as personality traits, mental abilities and behaviors – are largely controlled by genes. This implies that since we cannot change our genes, our life is predetermined. In other words, we are victims of our inheritance and we have little influence over our health.[39] In the media, this old belief is still propagated by some science journalists who portray genes as powerful entities that have the power to define personal identity, determine the course of human affairs and explain social problems.[40]

Fortunately, Epigenetics has shattered this erroneous belief. This relatively new scientific discipline studies environmental signals that modulate gene expression and activity. Genes are either active or inactive.[41] Signals that modulate gene expression arise from the internal environment (mental events, e.g., thoughts and emotions) or the external environment (e.g., diet, exercise and exposure to sunlight).

The discoveries in epigenetics reveal that our genetic makeup does not control us. Our thoughts, emotions, and beliefs are constantly actively influencing the expression of our genes. How is this possible? Biologist Bruce Lipton suggests that it is the biochemistry of our body that determines the nature of our environment at the cellular level. We know that the biochemistry of our blood is greatly influenced by the chemical messengers that are produced in the brain, and that our brain chemistry is directly related to the content of our mental activity. Lipton claims that our thoughts, emotions, and beliefs can thus affect the activity of our genes.[42] Hence, we can all be considered as "epigenetic engineers."

It seems that we possess the crucial ability to influence the expression of our genes. For example, we can consciously choose to generate comforting thoughts, maintain emancipating beliefs, and experience positive emotions. In order to achieve these goals, we can learn to meditate. In this regard, some studies show that when we meditate, our genes express themselves in a way that is beneficial to our health. One of these studies was led by Herbert Benson, an American cardiologist who founded the Institute for Mind/Body Medicine at Massachusetts General Hospital in Boston. Benson and his colleagues compared the gene expression of healthy men and women (control group members) with that of experts in mind-body practices. On average, expert participants had practiced yoga or meditation for nine years. A blood sample was collected from all participants in the study, and the genetic material was extracted from the blood. This material was then analyzed to determine the differences between the two groups of participants in terms of gene activity. In total, the researchers found that 2,209 genes were expressed differently in both groups. These genes – which play an important role in stress, inflammation and aging – are also known to be associated with good health. This means that the same genes, which have a beneficial effect on health if we practice a mind-body technique, can have a harmful effect on our health if we do not practice such a technique.[43]

Benson and his colleagues performed another study to determine if gene expression in the control group could be transformed by a mind-body approach. These participants followed an eight-week program involving relaxation, visualization and deep abdominal breathing. At the end of the program, a new blood sample was collected from the control group to assess how gene expression had changed in these participants. What the researchers discovered is remarkable: 1,561 genes had changed their expression and 433 genetic "signatures" were now comparable to those found in mind-body experts.[44]

Other epigenetic studies reveal that positive emotions increase the expression of the associated genes to the activity of NK cells (natural killers) that play a key role in the immune system. Moreover, it has recently been shown that three months of intensive meditation practice causes a 30% increase in telomerase, an enzyme that repairs and stretches telomeres, which are the DNA sequences at the end of chromosomes. These telomeres play an important role in cell division, and their length is an indication of cell longevity. This discovery suggests that meditation could slow the aging of our cells.[45]

## Chapter 2

# Perceptions Beyond Space and Time

*The greatest mystery of science is the nature of consciousness.*
Nick Herbert

## Remote Viewing

During the mid-1970s, right in the middle of the Cold War, the US military hierarchy tried to find ways to compete with the USSR. The military, together with government agencies like the Central Intelligence Agency (CIA) and the Defense Intelligence Agency (DIA), decided to create the Stargate Project. The goal of this top-secret project was to verify if "remote viewing" – a form of extrasensory perception (ESP) allowing a viewer (also called a "perceiver") to give details about a distant target in space (a place, a thing or a person) inaccessible to the normal senses – was an effective way to obtain sensitive information. The 20-year Stargate Project was subsidized to the tune of $20 million. Many of the experiments in this project were conducted at the Stanford Research Institute (SRI) under the supervision of physicists Russell Targ and Harold Puthoff.

One type of remote viewing experience that was developed at SRI is to send a "sender" to an unknown location and then ask the remote viewer to describe that location. The description is done verbally or through a drawing: the information generally collected consists of an impression of knowledge as well as sensory impressions (images, sounds, smells). The extent to which the descriptions correspond to the place is then evaluated by a group of judges. It is considered successful when the viewer's description matches the sender's location. In another type of protocol, the senders remain still: they only choose an image from a *National Geographic* photography bank.[46]

Over time, the protocols tested by the SRI research team were refined and conducted in a double-blind manner – neither the experimenters nor the participants knew the target.

The Stargate Project has provided information related to strategic enemy sites or weapons, political leaders, terrorist groups or drug trafficking operations. One of the star viewers of this research program was the late New York painter Ingo Swann, who trained the viewers at the CIA and intelligence services of the US Defense for 10 years. Swann also helped physicists Targ and Puthoff to design the initial remote vision protocol. During one session, he precisely described a ring around Jupiter. Scientists were unaware of this ring, which was confirmed years later when the Pioneer 10 Space Probe flew over this planet.

Researcher, author and documentary producer Stephan Schwartz is another pioneer of remote viewing. He is famous for his use of this mental capacity in the location and reconstruction of archaeological sites. Schwartz has been part of several intuitive archeology expeditions, especially on the shores of Grand Bahama where he found the Brig Leander, and on St. Ann's Bay in Jamaica where the caravel of Christopher Columbus was discovered. From 1978 to 1979, he directed the Alexandria Project whose objective was to measure the effectiveness and relevance of using remote viewing as a tool to assist archaeological exploration. Four targets were selected: the ancient city of Marea, the tomb of Alexander the Great, the Great Library and a site known in Antiquity as "the Hill of Numerous Passages."

The first phase of the project began in the United States in the summer of 1978. Eleven viewers were chosen to participate in the project. A questionnaire was sent to them, as well as a standard card where colors and names had been removed to avoid any visual disturbance. All data was sent to a bailiff in Los Angeles for analysis. Based on all the answers of the 11 viewers, it was possible to define three important research areas: one near the Nabi Daniel Mosque, another far east next to a kind of park, and the last on the main peninsula separating the western port from the eastern port.

The second phase of the project could now be launched. In March 1979, Stephan Schwartz and some members of his team, as well as two of the members of the original team, Hella Hamid and Georges McMullen, headed for Alexandria. With the help of the two viewers, the chosen areas were explored based on the data collected in the United States. The results were spectacular because the intuitive data made it possible to locate and describe several sites. These results include the discoveries of the palace of Mark Antony, the Ptolemaic palace of Cleopatra, the ruins of the lighthouse and the city of Marea. Moreover, the data from the viewers paved the way for the first modern cartography of the eastern port. According to Schwartz, the importance of the data obtained intuitively lies in the speed of the verification; divers can simply descend in the zones indicated in order to observe directly.[47]

Coming back to the Stargate Project, hundreds of experiments on remote viewing were conducted at SRI. When this research program was completed, the CIA asked for a report. This report aimed at determining whether the results of these experiments supported the idea that remote viewing can be useful in the collection of sensitive information. Jessica Utts, Statistics Professor at the University of California, Irvine, and Ray Hyman, Professor Emeritus of Psychology at the University of Oregon and notorious skeptic, were entrusted with the task of evaluating this research program. The conclusion of the report was mixed: while Utts stated categorically that proof of remote viewing was well established, unsurprisingly, Hyman declared that he did not agree with this conclusion. However, he acknowledged that the results of the Stargate experiments were not due to statistical errors or methodological problems.[48]

Other researchers have replicated the results of this program under strictly controlled conditions. For instance, Robert Jahn, Brenda Dunne, and their colleagues from the Princeton Engineering Anomalies Research (PEAR) Laboratory at Princeton University conducted 653 trials between 1976 and 1999, involving 72 participants. Several of these tests were conducted in a precognitive way – the future target was randomly chosen once the viewer had recorded their impressions. Statistical analysis of these 653 tests revealed that the probability of the results of these tests being due to chance was one in 33 million.[49]

## Mental Connections Beyond Space

Joseph Banks Rhine was a botanist who had also studied psychology. Interested in extrasensory perception (ESP – the term was coined by him) – which includes telepathy, clairvoyance, clairaudience and precognition – he decided in the 1930s to develop a statistical approach in order to study it scientifically. He collaborated with colleague, Karl Zener, at Duke University who was then Director of the Department of Psychology. Together, they created what would be called Zener cards, which include five types of symbols: a star, three undulating lines, a plus sign, a circle and a square. In total, the Zener card pack contains 25 cards, and consists of a repetition of five cards per symbol. In a purely random way, participants in Rhine's experiments have a one in five (or 20%) chance of guessing which symbols are selected. Results greater than 20% are considered to suggest the existence of ESP.

For decades, Rhine and his colleagues have used these cards to test various aspects of ESP. Hundreds of experiments involving thousands of trials were conducted. In one category of these experiments, to study telepathy, the pack of cards is carefully mixed. Then, one of the two participants, the sender, chooses a card and tries to mentally transmit the symbol appearing on the card to another participant, the receiver, which is at a distance. The perceiver must then correctly identify the mental "information" transmitted by the sender. The results obtained in these hundreds of experiments clearly point to the existence of the various categories of ESP.

While working with Zener cards, Rhine discovered that participants' psi performance weakens when experiments are repeated multiple times. This is hardly surprising since trying to guess for hundreds or even thousands of tests, whose symbols have been chosen, is extremely monotonous. Therefore, after Rhine, parapsychology researchers set out to design new experimental protocols to overcome this problem.[50]

In the 1970s, psychologist William Braud of the University of Houston, parapsychologist Charles Honorton of the Maimonides Medical Center in New York and psychologist Adrian Parker of the University of Edinburgh separately speculated that a weakening in sensory stimulation should promote the occurrence of ESP. This hypothesis was based on the observation that the reduction of mental "noise" – the constant and involuntary chatter of the mind – is associated with a greater manifestation of psi phenomena. This observation has as its origin the Vedas, the religious texts of ancient India. Braud, Honorton, and Parker posited that psi is a weak signal that is usually concealed by more intense signals from our ordinary senses that are constantly inundating us. Independently of each other, these researchers developed a telepathy test based on a sensory deprivation technique called the "Ganzfeld Experiment." This technique, which was initially developed to study visual mental imagery, quickly leads to a pleasant state of consciousness resembling reverie. According to the three researchers, blocking ordinary sensory stimulation *via* the Ganzfeld method should increase the ability to perceive subtle impressions.[51]

Experiments designed to test telepathy, and based on this procedure, are carried out in three phases: preparing, sending and judging. During the preparation phase, the receiver sits in a comfortable reclining chair in the Ganzfeld room. He has headphones on his ears and translucent half-spheres (usually ping-pong balls cut in half) on his eyes. He listens to a steady white noise through the headphones as a red light is diffused across his face. Then, the experimenter asks a research assistant to randomly select a "batch of targets" from a database that has multiple batches. Each batch contains four images, and all images are hidden in opaque envelopes.

During the sending phase, the experimenter gives the sender the target image. The sender must then attempt to mentally send the image to the receiver, who is in another room. To do this, it is recommended that senders "immerse themselves" in the target image and transmit everything that this image subjectively evokes. Finally, during the judgment phase, the sender and the receiver are informed that the sending phase is over. The white noise and the red lamp are turned off, and the ping-pong half-balls covering the receiver's eyes are removed. Then, the experimenter presents the receiver with copies of the four images that are part of the selected batch and asks him to rank them from 1 to 4 according to how it matches his subjective impressions felt during the sending phase. A "success" is achieved only if the receiver assigns the first place to the target, otherwise the test is considered a "failure." Statistically, the expected success rate is 25%, that is, with random responses, success should be achieved after every four trials.

At the beginning of the 1980s, Charles Honorton refined the methodological quality of Ganzfeld experiments by developing a fully automated protocol called "Autoganzfeld." In this procedure, the targets are composed of short audio-video clips. The batch with the decoys and the target clip is randomly selected by a computer program, and a closed-circuit video system presents the target to the receiver on a screen. On the other hand, the interactions between the experimenter, the receiver and the sender are fully automated, and at the end of the experiment, a computer randomly presents the target and the three decoys on a screen in the receiver room.[52]

The Autoganzfeld Program of Honorton and his colleagues lasted several years. A total of 240 people participated in 354 sessions as receivers, and the success rate was 37%. The probability of these results being due to chance was one in 45,000.[53] Following Charles Honorton's work, independent replication studies were successfully conducted by a number of researchers, including Adrian Parker (now at the University of Gothenburg in Sweden), Dick Bierman (University of Amsterdam), Daryl Bem (Cornell University) and Dean Radin (Institute of Noetic Science, California). A meta-analysis of the replication studies conducted by these researchers revealed an overall success rate of 33.2% with an associated probability greater than one million billion odds against 1.[54] Up to now, about 50 researchers from several countries have succeeded in reproducing these results, and other meta-analyses have confirmed the existence of telepathy when it is studied using the Autoganzfeld protocol.[55]

Some might argue that a 33% success rate (eight points above chance) is not so impressive. However, it should be noted that, in general, all psi effects are relatively small when data are obtained from participants with no particular abilities. Nevertheless, when special populations are tested, higher success rates are measured. For instance, a 47% success rate was reported when the Autoganzfeld protocol is tested with artists.[56] In addition, psi researchers have discovered that the more the sender is captivated by the video clips, the more the receiver is able to capture the target. This is not the case for static images, which produce lower scores.

## The Future in the Present

Research on remote viewing and telepathy shows that it is possible to acquire information independently of space without the use of the ordinary senses. Other psi research on presentiments – vague impressions that something is going to happen, but without being aware of what it might be – indicates that it is also possible to collect information about events that have not yet occurred.

There are several famous anecdotes about presentiments. Some of them are about Winston Churchill, who was Prime Minister of the United Kingdom from 1940 to 1945, and from 1951 to 1955. One evening, during the Second World War, Churchill invited three ministers of his government to dinner at 10 Downing Street. During the meal, a German raid began in London. Accustomed to sirens, Churchill and the three ministers continued to eat calmly. Then, all of a sudden, the Prime Minister got up from the table and went to the kitchen, where two of his housekeepers were working near a window. Churchill then ordered them to go down to the shelter. A few minutes later, a bomb crashed on the Prime Minister's residence, destroying the kitchen completely.

During the Second World War, Winston Churchill usually visited the anti-aircraft batteries that defended the British capital against German bombers. One night, after talking to the soldiers, he was about to leave them and headed for his vehicle. As usual, the right door at the back was open because this is where Churchill always sat. But this time, he decided against it and sat on the opposite side. The very next moment, a bomb exploded on the right side of the vehicle. The Prime Minister escaped unscathed. Later, he confided to his wife that a kind of inner voice whispered to him not to sit in his usual place.[57]

While presentiment is a vague impression, premonition is a belief that something will happen in the future. In some cases, a premonition manifests itself through dreams, known as premonitory dreams. A famous example of this type of dream was reported by David Booth, an office worker in Cincinnati, Ohio. For 10 nights in May 1979, Booth had the same nightmare:

he saw an American Airlines DC-10 plane hesitate to take off, and he heard the sound of the huge engines losing speed. He saw the plane nose up, then lose altitude before crashing to the ground. The plane became engulfed in flames right away – gigantic luminous flames, followed by thick black smoke. David Booth always woke up at that moment... On May 22, he called both American Airlines and the Federal Aviation Administration at Cincinnati International Airport. Four days later, an American Airlines DC-10 crashed over Chicago International Airport. Almost 300 people were killed – one of the worst air disasters in US history.[58]

Dr. Larry Dossey, who coined the term "non-local mind," proposed that presentiment, the perception of which sometimes seems irrational, relies on physical sensations – anxiety without apparent cause, unpleasant impressions – and is a warning system to protect us from danger. In this respect, Dossey cited the works of William Cox. In the 1950s, this researcher and businessman studied the visitor numbers in train accidents. He discovered that they usually had fewer passengers on board than the trains that reached their destination without any problem. Thus, on June 15, 1952, the day of its accident, the *Georgian* of the Chicago and Eastern Illinois Railroad had only nine travelers, while it carried an average of 62 people just days earlier.[59]

Experimental studies of precognition – the knowledge of information regarding future events – have been conducted for more than 80 years. In the first experiments, the participants had to guess which of the potential targets (for example, faces of a die, a colored bulb among several, or symbols on Zener cards) would subsequently be chosen randomly. A meta-analysis including more than 300 studies (published between 1935 and 1987) of precognition that used such targets has produced a probability of one out of $10^{25}$ (10 million billion)[60] that the results of these studies are due to chance. It therefore seems legitimate to exclude chance in the explanation of these results.

A few decades ago, Dean Radin conceived an experimental protocol to study presentiments. His hypothetical question was: are future emotional states (or future sensations) detectable in the current activity of the nervous system? In this protocol, participants sit on a chair in front of a computer screen. An electrode is attached to the middle finger of their left hand to measure their heart rate and the blood volume at the end of that finger. In addition, a sensor is attached to the thumb or forefinger of their left hand. This sensor records the fluctuations of the conductance of their skin, which is an index of emotional reactivity. Participants hold a computer mouse in their right hand and press the mouse button when they are ready to begin. Immediately, the computer randomly selects a target photo from a bank of 120 images. The computer screen remains blank. Then, after five seconds, the selected photo is displayed on the screen for three seconds. Then the screen turns white again (for 5 to 10 seconds). This step is followed by a rest period of five seconds. After this rest period, participants must press the mouse button again when they are ready for a retest.

During a session, 40 images were presented to participants. One of the categories of images (e.g., erotic images or autopsies) provoked strong emotional reactions, while the other category of images (such as pleasing images of landscapes and scenes of nature) was intended to provoke a state of serenity. The three types of physiological measures were averaged for all "emotional" and "calm" trials.

In keeping with what psychophysiological research on emotions has shown, seeing emotional images resulted in deceleration of heart rate, decreased blood volume, and increased skin conductance. However, what is fascinating is that the psychophysiological responses to emotional images started even before these images were presented on the screen, in other words, as soon as the mouse button was pressed. At the end of the experiment, participants were asked if they had been aware of the images that were going to be presented. Most said they were not aware. For Dean Radin, this indicates that presentiment and our reactions are largely an unconscious process. The unawareness of our reactions makes them possibly more effective, according to Radin, because it allows that to circumvent the psychological defenses likely to block psi perceptions.

In total, Dean Radin and his colleagues conducted four experiments on presentiment. The probability that the results of these experiments were due to chance was one in 125,000.[61] Radin's results were reproduced by psychology researcher Dick Bierman. In this replication study, participants were scanned using a functional magnetic resonance imaging (fMRI) device as they focused their attention on images projected by a computer. These images were chosen randomly. The results showed that specific regions of the brain involved in emotions were activated before the emotional images appeared.[62] These results indicate that the brains of the participants in this study responded to events that were not yet actualized.

Forty studies conducted in several different laboratories have made it possible to reproduce the results obtained by Radin and Bierman. These studies included physiological measures such as electroencephalography (EEG), the signal dependent on cerebral blood oxygen level (fMRI), heart rate, pupillary dilatation and skin conductance. A meta-analysis of these studies confirmed that we can sometimes unconsciously perceive our future.[63]

In 2011, Daryl Bem, Professor Emeritus and a recognized psychology researcher from Cornell University, published a study in the *Journal of Personality and Social Psychology* (JPSP) that included nine different protocols aimed at testing precognition.[64] In all, a thousand university students participated in the study. In one of the protocols, the following instructions were presented to participants:

*This is an experiment that aims to test ESP. It takes about 20 minutes and its progress is managed entirely by the computer. First, you will have to answer a few questions. Then, for each test, two curtains will appear on the screen, one next to the other. One of them hides an image; the other conceals an empty wall. Your task is to click on the curtain that seems to hide the image. This one will open to allow you to see if you have selected the right curtain. There will be a total of 36 trials. Many of the images have explicitly erotic content (for example, couples having consensual and non-violent sex).*

The purpose of these instructions was to make participants believe that they were participating in a clairvoyance experiment. But in fact, the images and their position (right or left) were determined randomly by the computer after the choice of participants. As the location of the images was chosen randomly, participants theoretically had a one in two chance of locating each erotic image. However, participants were able to locate erotic images 53.1% of the time (instead of 50%). These results were statistically significant. As for emotionally neutral images, the success rate did not deviate from what could be expected from chance.

The results of the studies reviewed in this section of Chapter 2 confirm that it is sometimes possible to gain access, consciously or otherwise, to information from the future. These results, which bring us to question the notion of time, suggest that certain areas of the human psyche can "travel" along the temporal continuum. About our conception of time, Einstein asserted that, "Time does not flow in one direction," and then, "The distinction between past, present and future is an illusion."[65] Maybe physicist Philippe Guillemant, a researcher at the CNRS, is right when he theorizes that our future is already realized in multiple versions that coexist simultaneously, but in a state of potentials not yet experienced.[66]

# Chapter 3

# When the Mind Influences Matter

*Not only is the Universe stranger than we imagine, but it is stranger than we can imagine.*
Arthur Eddington

The preceding chapters have shown that we can have access to information, beyond space and time, and without using our ordinary senses. Can one also mentally influence the physical world by using what parapsychologists (also named psi researchers) call psychokinesis (PK)?

Researchers have been looking at this fascinating issue for many decades. These psi researchers distinguish between macropsychokinesis (or macro-PK) and micropsychokinesis (or micro-PK). Macro-PK is a macro-physical psychokinesis, that is, it refers to objects or systems large enough for direct observation. As for micro-PK, it refers mainly to psychokinesis exerted on random microphysical systems (for instance, electronic noise).

Macro-PK can be revealed through the levitation and displacement of small objects (telekinesis), spoon bending, reshaping and bending of metal rods as well as through psychophotography (the mental impression of a photographic film). The experimental investigations of macro-PK have not always been carried out in a rigorous and systematic way. Furthermore, the results are ambiguous and certain subjects of experimentation have been outed as frauds. Given this, I will concentrate mainly on micro-PK in this chapter.

Dr. Joseph Banks Rhine was the first scientist to use statistical methods to study micro-PK. One day, during the 1930s, when JB Rhine was a professor at Duke University in Durham, North Carolina, a gambler paid him a visit. The latter told Rhine that he thought he was sometimes able to mentally influence the results of dice games. The belief in such an influence on dice was something widespread and old. JB Rhine realized that it would be easy and inexpensive to test this belief. As the preliminary results with the gambler in question proved promising, Rhine decided to carry out further research.[67]

This research program lasted for several years. Rhine tried to systematically eliminate the biases originating from unbalanced dice. Hence, the dice were first placed in special cups to prevent participants from manipulating their throw. The dice were then placed in power-driven rotating cages. During the experiments, the participants were asked to mentally "direct" the throw of the dice in order to unveil the face that had been pre-selected. A new face was chosen only after several throws.

In the early 1990s, parapsychology researchers Dean Radin and Diane Ferrari performed a meta-analysis of 73 micro-PK studies based on the protocol developed by Rhine.[68] These studies, carried out mainly in the 1940s and 1950s, had been conducted by 52 different researchers. The meta-analysis included 148 experiments covering 2.6 million dice throws. These experiments produced a relatively small but highly significant effect (probability of one in one billion). When there was no intention applied to the falling of the dice during the control condition, no significant effect was detected.

## Micropsychokinesis and Random Number Generators

Helmut Schmidt, a physicist working at the Boeing Laboratory in Seattle, created random number generators (RNGs) in the late 1950s to study the possibility that the human mind could interact directly with matter. Based on quantum processes, an RNG is a circuit that randomly produces peaks of electronic activity hundreds or even thousands of times per second. The electronic activity generated by the RNG is converted into random bit sequences, 0s and 1s. In a typical experiment, participants are asked to try to mentally influence the results of the RNG so that this device generates more 0s than 1s (or vice versa). During the control condition, participants do not mentally influence the activity of the RNG.

Typically, participants perform hundreds of rounds of tests, and they can receive feedback on the distribution of random events in the form of sounds or graphics. Contemporary RNGs are small devices connected to a computer *via* a USB port. All aspects of RNG experiments – including the presentation of instructions, the feedback provided on a trial-and-test basis, and the recording and analysis of data – are now automated.

Over the last 50 years, several hundred RNG studies have been carried out. In 1989, Dean Radin and psychology researcher Roger Nelson of Princeton University worked on a meta-analysis of studies conducted between 1959 and 1987. This statistical analysis included 597 studies that had been conducted by 68 different researchers. Meta-analysis of these 653 tests revealed that the probability of the results of these tests being due to chance was one in 33 million.[69]

More recently, engineer Robert G. Jahn and his colleagues at the PEAR Laboratory published a meta-analysis based on twelve years of RNG experiments conducted by their research group.[70] These experiments involved more than 100 participants, who had been instructed to attempt to intentionally influence the data produced by an RNG by either moving them up the hazard curve (the "high" condition), or below the hazard curve (the "low" condition). The results showed that the RNG data was higher when participants wanted high scores, while the data dropped more when participants wanted low scores. Even if the effects recorded were not very significant, all the results produced a probability of 35 billion against one. In certain PEAR experiments, participants were thousands of kilometers away from the RNG. Nevertheless, there was no decrease related to distance.

Robert Jahn and his colleagues also conducted macro-PK "placement" studies. These studies aimed to determine whether human intention can affect the movement of macroscopic objects. To achieve this, the researchers used a random cascade mechanism. It was actually a 10-foot high experimental device that allowed 9,000 polystyrene balls to fall through a matrix of 330 tees. The balls were distributed into 19 boxes. A photoelectric process was used to count these balls as they entered the boxes. Participants were asked to focus and try to mentally shift the distribution of the balls to the right or left, relative to a normal control distribution. Overall, more than 300 experimental sessions were conducted with about 20 different participants, and slightly significant results were obtained.[71]

## Collective Attention Effects

Micro-PK studies indicate that when an individual directs his attention and intent to random physical processes, such as the falling of a die or the "behavior" of an RNG, these processes no longer operate totally randomly. In this context, it is legitimate to ask the following question: what happens to the micro-PK effect when several individuals are involved?

In the mid-1990s, while still a PEAR affiliate, Roger Nelson decided to explore this important issue using RNGs. His work, and subsequent replication studies by other parapsychology researchers, suggest the existence of a "collective attention effect." Such an effect could be linked to the creation of "fields of consciousness" (these fields are designed in a similar way to the classical concept of field in physics).

A field of cosmic consciousness would permeate space and time. This is quite an ancient concept, first put forward by the Vedic sages (rishis) of ancient India who posited that there is a field of pure consciousness or universal Self, which they called Brahman. This field of cosmic consciousness would encompass all individual human consciousness.

As Dean Radin points out[72] the notion of field of consciousness is related to the subjective impression that individual thoughts, which are aligned and coherent, seem to merge into a group thought. This would occur when a group of individuals are on the same "wavelength," for example, during religious rituals or while visiting sacred places. A field of consciousness could also be created when a group of people sing in chorus or attend a moving speech. In such moments, it is as though the normally dispersed attention of group members becomes focused and a collective mental coherence sets in. In such moments, microphysical changes can apparently be detected in the local environment.

So far, more than a hundred studies of collective attention effects have been conducted using RNGs. In these studies, RNG activity is measured continuously, while groups of individuals are involved in stimulating activities. It should be noted that the experimental protocol of these studies does not imply that participants must specifically focus their intention, or direct their attention, towards the RNG. In terms of data analysis, scientists are looking for significant deviations from the sequences of numbers normally generated randomly. Thus, in a sequence of 1,000 bits, generated in a purely random manner, and without the RNG being influenced in any way, 500 0s and 500 1s will be obtained (with a perfectly calibrated RNG). With a group awareness effect, there should be more 0s or more 1s than expected due to chance alone.

The probability that the results of these studies are due to chance is one chance in more than one million: The effect of collective attention does seem to exist. However, the results of these studies vary widely. This is likely due to differences in group composition, activities by members of these groups, and environmental contexts.[73] Dean Radin proposed that this inter-study variability could also be related to the difficulty of determining exactly when collective mental coherence appears.

To explore this possibility, he worked with Skip Atwater, former Director of Research at the Monroe Institute in Virginia. Radin and Atwater wanted to specifically investigate whether the binaural sounds of the Hemi-Sync method, which was developed by this private institute, can promote group awareness effect.[74]

When listening to two slightly different sonic frequencies with stereo headphones, for example, 500 Hz in the right ear and 495 Hz in the left ear, the neurons of the auditory system process these sounds so that the individual with the earphones hears a third sound that results from the difference between the two frequencies, or 5 Hz. It is this differential frequency that constitutes what is called a binaural sound.[75] Research at the Monroe Institute indicates that binaural sounds induce brainwave synchronization in both hemispheres. The Hemi-Sync method promotes, among other things, relaxation and stress reduction.

The research by Dean Radin and Skip Atwater was conducted during the *Gateway Voyage*, an intensive six-day program at the Monroe Institute. The purpose of this program is to enable participants to explore and develop different levels of consciousness. During this program, for several hours daily, the participants – usually about twenty – listen separately, each in their room, to various binaural sounds associated with the Hemi-Sync method. After listening to these sounds individually (with stereo earphones), participants gather for 45 minutes to discuss their experiences. The cycle of sound listening-discussion is repeated several times a day during the entire duration of the program.

Radin and Atwater speculated that listening to binaural sounds should lead to the same type of brain rhythms in each participant, and that this could produce collective mental coherence. While assuming that the collective attention effect does exist, Radin and Atwater also hypothesized that such an effect should be more apparent during binaural listening sessions than during meals or when participants sleep. In order to test this hypothesis, these researchers performed correlation analyses of the RNG data collected over 14 weekends. Most program participants were unaware of the application of RNG during the research. For the control condition, RNG data were collected during eight weekends, while there was no program at the Monroe Institute.

In all, 108 billion random bits were produced and stored. Correlation analyses confirmed the hypothesis of the two researchers, i.e. a statistically significant effect was measured during the programs (probability that the results were random: one chance out of 500), while no significant effect was noted during the control condition. In addition, data produced by an RNG found at a distance from the location of the programs showed no effect.[76]

Studies by Nelson, Radin, and other researchers in the field of psi suggest that some factors influence the magnitude of collective attention effects. First, collective attention and collective mental coherence would be greater when the attention of group members is focused on the same object or event. The number of individuals in the group would be another factor influencing the magnitude of such effects.[77]

## Masaru Emoto and the Sensitivity of Water

Masaru Emoto, who died in 2014, was a Japanese author and internationally known independent researcher for his theory of the effects of thought and emotions on water. After studying international relations, Emoto obtained a PhD in Alternative Medicine from Yokohama University. Fascinated by water, Masaru Emoto was also President of *Project of Love and Thanks to Water*.

Emoto developed, with the help of some collaborators, a method of observing water crystals frozen by photography. In this method, crystallization is achieved by freezing water samples in Petri dishes for three hours. Droplets of ice subsequently form on the surface, and the water crystals appear under a projection of light. It is then that a photograph of the crystals is taken at a fast speed. With this method, Emoto discovered that the crystal structure varies greatly depending on the source of the water. Thus, the pure and vivid waters form breathtaking and harmonious crystals, while the stagnant or waste waters do not form crystals and if they do happen to form crystals, they are incomplete and disharmonious.

Using this method, Masaru Emoto tested the sensitivity of water against the "energy" emitted by thought and emotions, speech, image and music. His work revealed that crystals react by structural changes to different influences. Thus, even simple words written on paper (for example, "hate" and "love") presented to water change the shape of the crystals in a specific way. In general, the tests conducted by Emoto and his colleagues revealed that beautiful music (for example, that of Johann Sebastian Bach), as well as thoughts, emotions and positive words lead to the formation of beautiful crystals. On the other hand, negative thoughts, emotions and words, as well as hard rock, lead to the formation of deformed and aesthetically unattractive crystals.

The results of the work of Emoto and his team have been published in a few books showing photographs of various water crystals in a variety of situations.[78] Emoto and his work became famous after his experiments on water molecules were included in the film *What the Bleep Do We Know!?*

Later on, Masaru Emoto, who did not have a traditional scientific background (some skeptics did not hesitate to call his discovery a sham), began to collaborate with the highly regarded Dean Radin. The latter made sure to eliminate the main methodological weaknesses that characterized Emoto's preliminary work. Thus, the possible waiting effects on the part of the photographer, the judges and the data analyst were controlled by using a blind procedure. In a first study,[79] a group of about 2,000 people in Tokyo was asked to focus on the idea of having positive intentions toward water samples in an electromagnetically shielded room at the Institute of Noetic Sciences in Petaluma, California. Participants were unaware that there were identical water samples that had been placed elsewhere in this institute.

The ice crystals formed from the two sets of water samples were blind photographed. In addition, the images obtained were blindly evaluated for aesthetic appearance by 100 independent judges. Ice crystals formed from "treated" water were judged to be significantly more beautiful by the judges.

In a subsequent replication study by Radin, Emoto and their colleagues[80] 1,900 people in Austria and Germany sent their intention to two bottles of water in the electromagnetically shielded room at the Institute for Noetic Sciences. As in the previous study, two control bottles had been placed elsewhere at the institute. After the water was frozen, crystallized, and the photos were taken, 2,500 independent judges blindly assessed the beauty of the crystals. A similar result was obtained – the treated water crystals were found to be more beautiful – symmetrical, well formed, and attractive – than the crystals formed from the control water.

Since the human body is 70% water, it is easy to understand why the implications of these studies for our health can be important.

## The Observer Effect in the Quantum World

The micro-PK studies conducted with RNGs suggest that it is possible to mentally influence processes taking place in the microphysical world. One of the biggest oddities of this world is the fact that a quantum object behaves like a wave indefinitely until it is observed. As if the quantum object in question knew that it is being observed. This quirk is indicated by a double-slit interferometer. It is a device that sends particles of light (or photons, but this is also true of electrons and elementary particles) through a wall pierced with two small parallel slits and records the patterns of light that appear on a screen behind the slits. Insofar as the photons behave like tiny, separate beads, the patterns of light should always feature two bands of light. This is what is actually recorded if we follow each photon as they pass through the slits. However, if we do not follow the path of the photons, we observe a diffraction pattern, an area where dark and illuminated fringes alternate (called interference fringes). Such a pattern illustrates the wave-like character of light. It thus has a double nature: it is both wave and corpuscle. And how it manifests itself depends on how one seeks to apprehend it. This mysterious phenomenon is called by physicists the "problem of measurement."[81]

There are other QM concepts that are closely associated with this phenomenon. The "wave packet reduction" (or "wave function collapse") is one of these concepts. According to QM, a physical system has multiple possible states but physically not manifested – these states are said to be superimposed — when this system is not observed and measured, disturbed by an observation/measurement. When an observation/measurement is made, the superimposed states must collapse into one and only one possibility. In other words, the observation not only disturbs what is measured, it also produces the effect.[82]

The problem of measurement defies common sense since it calls into question the assumption that there is an objective reality totally independent of the observers. This problem has deeply shaken what physicists believed about the physical world. To illustrate the strange and problematic nature of this phenomenon, Erwin Schrödinger, one of the fathers of the QM, imagined in 1935 a famous thought experiment. This thought experiment is called "Schrödinger's Cat." In this famous experiment, we imagine a cat locked in a box with a deadly device. This device is designed to release cyanide if a radioactivity detector detects the disintegration of a radioactive atom. Since an atom has a one in two chance of disintegrating over a period of one minute, the probabilistic equations of the QM describe the fate of the cat as 50% alive and 50% dead. These equations thus illustrate a superposition of two states. If an observer opens the box, they decrease their ignorance about the cat's state of health – which corresponds to the collapse of the wave function – and discovers whether the latter is 100% alive or 100% dead.[83]

The problem of measurement and the effect of the observer have led several famous physicists – such as Bohr, Pauli, Schrödinger and von Neumann – to think that consciousness plays an important role in shaping the physical world. Most physicists reject this idea and opt for other theoretical avenues to explain the problem of measurement. However, some theoretical physicists have gone even further than the fathers of QM. Thus, John Wheeler proposed the "Participatory Anthropic Principle." According to this principle, the presence of observers is necessary for the Universe to exist. In other words, consciousness must have existed before the appearance of the physical world.[84]

To learn more about the role of consciousness in the collapse of the wave function in QM, Dean Radin and his colleagues used a double-slit optical system a few years back.[85] These researchers hypothesized that the intensity of the interference pattern produced by this optical system would decrease when participants in the study would focus their attention on the double slit. A total of 137 volunteers took part in this study and 250 test sessions were conducted. Each session consisted of 40 trials (varying between 15 and 30 seconds). For each of these trials, participants were asked to focus either on the double slit (the experimental condition), or away from the double slit (the control condition). The data collected revealed that, on average, the intensity of the interference figure decreased significantly in the expected direction (probability of 184,000 against one). In addition, the researchers conducted 250 control sessions without any observer. These sessions were intended to detect the presence of potential artifacts (for instance, related to equipment and methods of analysis). No effect was detected. These results, which were published in the journal *Physics Essays* in 2012, support the interpretation of QM according to which consciousness is related to the problem of measurement. In addition, Radin and his colleagues found that factors such as experience in meditation, the degree of focus of attention, and open-mindedness were correlated in a predictable direction with the intensity disturbances of the interference figure.

More recently, Radin and colleagues conducted another study using a similar double-slit optical system and experimental protocol.[86] The objective of this new study was to determine whether the "non-physical" interaction proposed by some physicists is better interpreted as an active influence of consciousness or as a passive form of observation. This time, the experiment was conducted on the Internet by sending the data of the double-slit system to the web browsers of the participants. In all, 1,479 people from 77 countries participated in 2,985 experimental sessions. These participants had to either focus their attention or not on a signal that represented in real time the level of interference produced by the optical system. The results were similar to those of the previous study. Remarkably, these results also showed that the direction of the effect, which was consistent with the intent of the observers, was not affected by the distance between the participants and the double-slit system (this distance ranged from 1 km to 18,000 km). On the other hand, no effect was observed in more than 5,000 control sessions recorded by a computer simulating a human participant.

## Is the Brain a Quantum System?

In the 1930s, John von Neumann, the Hungarian-American mathematician and physicist, was the first to propose that in QM the observer of a measurement is not only the instrument that is used but also the consciousness of the physicist who notes this measurement. This idea was taken up decades later, first by Eugene Wigner, followed by Henry Stapp, a theoretical physicist working at the University of California, Berkeley. Stapp developed a model postulating quantum effects in neuronal ion channels. When these channels open, the neurotransmitters are released into the synaptic cleft. Since synapses enable the creation of neuronal assemblies, and these structures are closely related to mental processes, Stapp proposes that these structures can trigger quantum effects in ion channels. Considering that the diameter of these channels is one nanometer ($10^{-9}$ meter), this physicist also proposes that these channels belong to the microphysical world, and that their activity is governed by the rules of QM. Furthermore, Stapp argues that a goal-oriented attention effort directly modulates the reduction of the wave packet associated with the activity of the ion channels. This would be how free will is implemented in the brain.[87]

There are other models of mind-brain interactions that are based on QM. For example, neuroscientist John Eccles and physicist Friedrich Beck suggested that the probability of release of chemical messengers from presynaptic vesicles – the small membrane vesicles at the synaptic junction of neurons that contain chemical messengers – is regulated by quantum processes. This hypothesis is based on the fact that the extremely small size of the synaptic vesicles is comparable to that of other structures and phenomena studied by QM. Furthermore, Eccles proposed that dendrites – the branch-like extensions of a neuron that conduct electrical impulses from adjacent neurons to the cell body – act as the basic receptor units of the cerebral cortex (the mantle of grey matter covering the surface of the cerebral hemispheres). Sets of dendrites (or "dendrons") are linked to units of mental experience (or "psychons"). During intentional and voluntary thoughts, psychons would act on the dendrons by temporarily increasing the likelihood of release of neurotransmitters *via* presynaptic vesicles.[88]

Neither Stapp, nor Eccles nor Beck attempted to explain the mind and consciousness in quantum terms, unlike the English physicist and mathematician Roger Penrose and the American anesthesiologist Stuart Hameroff. Penrose does not believe in the concept of wave function collapse. His proposal is to replace this QM concept with that of "objective reduction." The latter would be of a gravitational and non-local nature – it would be based on remote effects and would make it possible to establish a quantum coherence between spatially distant objects. According to Penrose and Hameroff, objective reduction would occur at the microtubule level in the brain. These are tiny tubes (about 25 nanometers in diameter) made of spirally wound proteins (called tubulin). Microtubules serve as a structural support for neurons as well as for all the cells of the body. Tubulin molecules can be either in an elongated state or in a contracted state. For Penrose and Hameroff, a superposition of quantum states would explain this phenomenon. On the other hand, the very small size of microtubules, their spiral protein structure and the fact that these organelles are relatively isolated from the external environment provide all the essential conditions for promoting quantum coherence.[89]

According to Penrose and Hameroff, their model would explain the unity of consciousness and the origin of free will. Philosophers Rick Grush and Patricia Churchland strongly disagree and have raised several objections that are difficult to circumvent. First, microtubules are found not only in the human brain, but also in all animal and plant cells. Then, several anesthetic substances act without these organelles being affected and the destruction of microtubules does not lead to major effects on the consciousness. Finally, there is no evidence that these organelles are involved in central changes in states of consciousness such as the wake-sleep cycle.[90] I would add that the Penrose-Hameroff model is another attempt to reduce the mind to physical processes. An attempt that has failed. Instead of the electrochemical activity in neurons or neuronal assemblies, it is alleged quantum processes in microtubules that would be the source of mental activity. It is a model that is even more reductionist than most materialist theories, which are content to reduce the mind to cellular or molecular levels.

## A Device Supporting Intention?

William Tiller is Professor Emeritus of Stanford University's Department of Materials Science and Engineering. He has also been director of that department. During his long scientific career, which focused mainly on crystallization mechanisms, Tiller published some academic books and more than 200 articles.

This scientist did not feel entirely fulfilled by his official work at the university because he had the impression that his work only touched on a very narrow portion of reality. In parallel with his formal research, William Tiller decided in the 1970s to undertake a research program on the effects of thought on the material world. Eventually, he developed a device that allows us, he says, to record and implement intentions. It is a small electronic device that plugs into an electrical outlet and is placed near the material target that you want to influence. At first, you have to "load" the device with a specific intention. Tiller discovered that this process was effective when a small group of people entered a meditative state and focused for about fifteen minutes on the intention in question.

Using such an experimental protocol, William Tiller did several series of experiments that aimed to change the pH level of water. The pH makes it possible to estimate the concentration of hydrogen ions in an aqueous solution and measure the acidic or basic nature of this solution. The more acidic the solution, the lower the pH value and *vice versa*. The purpose of the first experiment was to increase the pH of 1. A few months after switching on the intention device, Tiller observed that the pH of the water had risen by 1. For the second experiment, the goal was to lower the pH of the water by 1. This goal was also achieved. The results of these experiments were published. Afterwards, these results were reproduced by research groups in London, Milan and the University of Florida.

Physicists are familiar with atoms, protons, electrons, and the four fundamental forces present in nature: gravitation, electromagnetic interaction, strong nuclear interaction, and weak nuclear interaction. However, to understand the results of his experiments with the intention host device, William Tiller believed that it is necessary to go beyond QM, the fundamental forces and the physical world known today, by imagining the existence of subtle energies and forces which still evade science.[91]

**Chapter 4**

# The Non-Local Influence of Intention on the Living

*There are only two ways to live your life. One is as though nothing is a miracle. The other is as though everything is a miracle.*
Albert Einstein

The studies presented in Chapter 3 demonstrate that we can influence the microphysical world non-locally. Other studies, which we will examine in this chapter, indicate that the human mind is also capable of bio-psychokinesis (bio-PK) – a non-local action on organic matter and living things.

The scientific investigation of bio-PK began in the 1960s. One of the pioneers in this field of research, Cleve Backster, was not a researcher, but rather a lie detector (or polygraph) specialist who worked for the CIA. One morning in February 1966, during a coffee break, this former member of the US Army Counterintelligence Service decided to water the plants in his office. Among them was a dragon tree – a plant with a very long stem. Inquisitive, Backster wondered, as he filled the watering can, how long it would take the water to travel from the roots to the leaves of the dragon tree. He decided to find out. For this purpose, he inserted one of the leaves of the plant between the two electrodes of a polygraph, then he tied the whole thing up with an elastic band. When the water reached that part of the leaf between the electrodes, the humidity would rise, leading to a drop in the electrical resistance of the leaf. This decrease would result in a rise in the outline of the printed paper roll of the polygraph.

Contrary to what Backster had anticipated after pouring the water, the curve slumped, similar to when a person experiences the fear of being caught in their lie. The CIA polygraphist wondered at that point whether a plant could react emotionally. If that were true, he thought, threatening the well-being of the plant would probably heighten the reaction of the outline. Backster decided to use a match to burn the leaf attached to the electrodes. Immediately, the plotter pen of the drop-down paper of the polygraph shot up. It was as if the plant had been terrified after perceiving Backster's menacing thought. The polygraph expert then went to his secretary's office to pick up matches. The polygraph still showed the alarm response of the plant when Backster returned to his office. He then lit a match and brought it close to the leaf – the tracer pen continued to show the alarm response. The polygraph expert then returned the matches to his secretary and the line gradually normalized.

For Cleve Backster, the extraordinary phenomenon he had just observed proved that plants possess some kind of psi perception. He named it "primary perception." This type of perception would make plants receptive to the thoughts and intentions of human beings. During the following decades, Backster performed a significant number of experiments that confirmed the existence of this primary perception in plants.[92]

Almost at the same time as Cleve Backster made his discovery, a chemist named Robert Miller performed an experiment that suggested that human intent could affect plants, not just in their immediate environment but also remotely. Miller was conducting research to determine the extent to which lighting variations affect the growth rate of rye. To carry out this investigation, Robert Miller used a device to measure the growth rate of this plant with an accuracy of thousandths of centimeters per hour. When lighting, temperature and irrigation were kept constant, the growth rate of the rye was relatively stable (0.0152 cm per hour). In 1967, Miller asked the spiritual healer Olga Worrall and her husband Ambrose – who treated their patients remotely through prayer – to try to mentally influence the growth of a single rye shoot, from their home, which was located 800 kilometers away from Robert Miller's laboratory. They settled on a date. On that day, at 9 o'clock in the evening, the couple visualized the rye shoot growing fiercely, enveloped in white light. The following day, Miller discovered that before the Worralls began to "pray" for the plant, the pattern produced by the device showed a slight slope that represented a normal growth rate. However, this outline began to deviate upwards at exactly 9 o'clock the previous evening, and at around 8 o'clock the next morning the growth rate was 0.1344 centimeters per hour, an increase of more than 800%.[93]

Following his chance discovery, Cleve Backster conducted many experiments demonstrating that cut or crushed leaves, as well as different organic materials – for example, eggs, yogurt and smear test of the palate – react to human intentions and emotions. Some of these experiments indicated that oral leukocytes (white blood cells from the mouth of a person) placed in a test tube react electrochemically to the emotional states of the donor, even if the donor is in a distant room, another building, or even another country.[94]

Similarly, physicist William Tiller and his colleagues used the device to record and implement intentions, and the protocol described in the previous chapter, to non-locally affect various biological processes. One of the experiments targeted the activity of enzymes, those proteins that catalyze chemical reactions in cells. Human liver enzymes were then harvested. The results showed a 30% increase in the level of activity of these enzymes. The purpose of another experiment was to measure the non-local impact of intention on the rate of development and maturation of fly maggots. In this experiment, the thoughts of the meditators were to increase the amount of adenosine triphosphate (or ATP) – the main source of energy at the cellular level – in maggot cells. Tiller and colleagues measured a 15% increase in ATP levels in these cells. This increase was associated with a shorter maggot development time, by almost 25%.[95]

Jean Barry, a French doctor from Bordeaux who had also been president of The Institut Métapsychique International (IMI), is another pioneering researcher on bio-PK. Barry tried to determine if it is possible for humans to mentally inhibit the growth of a fungus that can cause an infectious disease. In this experiment, the genetic purity of the fungus, which was grown in Petri dishes (a laboratory incubator), was controlled. The environmental conditions (temperature, humidity, and lighting) were identical for the boxes chosen to be influenced at random and for the boxes of the control group. Ten individuals were part of the experiment. During each session, five experimental boxes and five control boxes were assigned to each participant – they were seated five feet away from these boxes. They were asked to only focus their attention and intention on the five experimental boxes for fifteen minutes. At the end of each session, the boxes were weighed and compared. If the experimental boxes were lighter than the control boxes, a success was recorded, and if they were heavier, it was a failure. For 151 of the 195 sessions performed as part of this experiment, the growth of fungus in the experimental dishes was significantly lower than that of the control boxes.[96]

Another interesting bio-PK study was led by psychologist William Braud while he was working at the Mind Science Foundation in San Antonio, Texas.[97] In this study, during every session, Braud asked participants to visualize the slowing down of the rate of hemolysis (destruction) of red blood cells in human blood, found in test tubes in a remote room. Nine out of 32 participants managed to significantly reduce the hemolysis rate of red blood cells in the tubes. The results also revealed that this effect was more pronounced when it came to their own blood.

Other studies have tested the ability of individuals to modify the physiological activity of an animal. For example, researchers at the Institute for Parapsychology, in North Carolina, have examined bio-PK by assessing how long anesthetized mice took to wake up.[98] The mice were first put in pairs. The two members of the same pair were of the same sex, of the same mother, and they were of similar size. One mouse was selected to be part of the experimental group, while the other belonged to the control group. The mice in both groups were anesthetized with ether, and the study participants only had to focus on mice in the experimental group. Two series of tests were carried out. In the first one, the participants were close to the mice but could not touch them. In the second set of tests, the participants were in another room and observed the mice in the experimental group through a one-way mirror. The results were significant for both sets of trials: the treated mice woke up faster than the mice in the control group.

## Non-Local Effect of Intention on Others

Other studies have sought to evaluate the non-local effect of sender intent on the physiological activity of a person (the receiver) located at a distance – this experimental design is called a transceiver protocol. In these studies, the receiver, who is asked to remain calm and relaxed, is unaware for how long the sender "is sending" an intent, and at what moments he is doing so. Charles Tart, a psychology researcher at the University of California in Davis, is one of the pioneers of this type of study. In the early 1960s, Tart used such a protocol and acted as a transmitter to test whether remote receivers could also react to random electrical shocks he received. The skin conductance, heart rate, and blood volume of the receptors were measured. These physiological evaluations revealed that the receptors reacted significantly to the shocks Tart was subjected to. However, these receivers were completely oblivious of these events. Following the study by Charles Tart, Jean Barry (in France), engineer Eric Douglas Dean (Newark College of Engineering in New Jersey), and psychologist Erlendur Haraldsson (in Iceland) separately recorded significant changes in blood volume measured at the receiver's finger at the moment when transmitters, in some cases thousands of kilometers away, mentally directed their emotional states towards them.[99]

Psychology researcher William Braud has conducted several studies in the field of bio-PK, and he developed an innovative experimental procedure that he called "allobiofeedback" (from the Greek *allos*, meaning "other"). The traditional approach to biofeedback is to provide the individual who is being examined with information about a specific physiological measure, such as his heart rate. In experiments involving allobiofeedback, the information is provided to the transmitters to enable them to determine the extent to which they can mentally influence the physiology of the receivers located at a distance (in another room). In several of the experiments conducted by Braud and his colleagues, the physiological target selected was skin conductance (or electrodermal response), which reflects the activity of the sympathetic nervous system. A high electrodermal response indicates anxiety or stress. In these experiments, the transmitter observed, on a polygraph, the fluctuations of this response in the receiver, as they attempted to increase or decrease this physiological activity. The influence periods lasted typically for 30 seconds. The target physiological activity during these periods was compared to that measured during control periods without influence. The order of the influence periods and control periods was randomly determined, and the transmitters received the instructions *via* a headset.

The numerous experiments carried out by Braud and his colleagues, using this experimental approach – more than 300 sessions involving 62 agents and 270 receivers – indicated a remote mental influence on skin conductance during periods of influence. In one of these experiments, a first group of receivers had a normal level of skin conductance, while a second group of receptors showed an abnormally high level of skin conductance. The aim of the experiment was to reduce this physiological activity in both groups of receivers. The results showed a more marked reduction in skin conductance in the receivers from the second group.[100]

In other experiments conducted by Braud and his research team, the allobiofeedback was not used. Instead, senders were asked to immerse themselves in emotionally-charged images, as well as the desire of the intention to reach the receiver; or to imagine that the receiver was in a state corresponding to the intention issued. This different approach has also produced significant results.

William Braud and his long-time colleague, anthropologist Marilyn Schlitz, conducted a meta-analysis of the 37 experiments which they carried out until the early 1990s on bio-PK. These experiments, which involved measuring different physiological responses, included 655 sessions with 153 people acting as transmitters and 449 people (or animals) acting as receivers. The combined results of these various experiments generated a probability of one hundred billion against one.[101]

Researchers Deborah Delanoy and Sunita Sah from the University of Edinburgh conducted a variant of the experiment popularized by Braud. The purpose of their study was to verify if the receiver can react physiologically to an emotional state experienced by the transmitter. At first, the senders had to identify personal memories capable of inducing a positive and happy state. The senders also had to choose four objects that were emotionally neutral for them. During the experiment, a random process informed senders of sending periods, and assigned them as targets in either a "positive" state or a "neutral" state. For each sending period, the receivers – who were placed in a room 80 feet from that of the senders – had to try to guess the emotional state of the latter (positive or neutral). The skin conductance of the receivers was also measured.

Analysis of the results of the 32 sender/receiver pairs revealed that the receivers showed no particular ability to consciously perceive the emotional states of the senders. However, the skin conductance in the receivers was much more significant during the positive periods. Delanoy and Sah believe that this increase could be an index of unconscious psi reception.[102]

Transpersonal psychologist Jeanne Achterberg, a pioneer in the use of visualization in the healing process, conducted a functional magnetic resonance imaging (fMRI) study to determine what occurs in receivers' brains when healers direct positive intentions towards them. Eleven healers from Hawaii were paired with 11 receivers. These healers adhered to various therapeutic approaches and were far away when the scans were performed. As with the studies previously analyzed, healers had to connect with the receivers only during certain periods. During the control periods, they did not emit any intention. The results of the study showed significant differences between the two conditions. As a result, when the intention was sent, there was a significant increase in the level of oxygenation of the blood connected to neuronal activity in different regions of the receptors' brains. This suggests that it is actually possible for healers to connect with people from a distance, and to influence their brain activity.[103]

## Can Intention Influence DNA?

When we experience negative emotions or face stressful situations, our heartbeat becomes irregular. Researchers from the HeartMath Institute in California have shown that in such situations, when we learn to breathe calmly, the variations of our heart rate decrease as well as the level of stress. We then experience what is called "heart coherence." This physiological state, which provides a deep sense of well-being, is associated with a resonance between the rhythm of the heart and that of breathing.[104] Glen Rein is a biochemist who was affiliated to the

HeartMath Institute for a long time. When he worked there, Rein wondered if individuals achieving high cardiac coherence would be able to modify the structure of DNA through intention.

This biological macromolecule consists of four basic molecules: adenine (A), thymine (T), guanine (G), and cytosine (C). The sequence of these molecules, which work in pairs (A-T and C-G), is what carries the code of life. Recently, a team of American researchers working at the Scripps Institute, in La Jolla, managed to implant a third pair of artificial molecules (different from A-T and C-G) in the DNA of the bacterium, *Escherichia coli*. What is even more remarkable is that this bacterium was able to replicate the semi-synthetic DNA. This major discovery could eventually lead to huge breakthroughs. Floyd Romesberg, the director of this research team, said on this subject: "If you read a book written with four letters, you will not be able to tell very interesting stories. If you are given more letters, you can invent new words [...] and you will probably be able to tell more interesting stories."[105]

This suggests that living organisms on Earth could have evolved quite differently with DNA consisting of more than four letters. In this context, the discovery of Romesberg and colleagues suggests the possibility of improving the human species, for example by making it more resistant and increasing its longevity.

Romesberg and his team have used sophisticated genetic engineering techniques to complicate the DNA of a bacterium. However, is it possible to change the helical structure of this macromolecule with only the mind and intention serving as tools?

In order to answer this exciting and daring question, Glen Rein collected DNA samples that were placed in a beaker in front of study participants. They mentally tried to change the structure of the helix. Rein discovered that the higher the degree of cardiac coherence in the participants, the greater the structural changes in the DNA (twisting and untwisting its helix). To achieve a high degree of cardiac coherence, many of the participants generated a strong sense of unconditional love.[106]

Glen Rein also conducted a study with Leonard Laskow, a medical doctor who also works as a healer. In his healing work, Laskow combines unconditional love, directed intention, and mental imagery. With this approach, Leonard Laskow has demonstrated that he is able to slow the growth of bacteria or protect bacteria that are subject to doses of antibiotics.

In this study, Rein prepared five Petri dishes, each containing an identical number of cancer cells, and asked Laskow to make a different intention for each of these boxes while holding them in his hand. Leonard Laskow's task was to inhibit the growth of cancer cells by using different intentions and strategies. The growth rate of the malignant cells was measured by evaluating the amount of radioactive thymidine that they absorbed.

The healer began the experimentation with meditation to enter a state of unconditional love that would foster connection with the cancer cells. The five different intentions and strategies he used were: 1. To allow a return to the harmony and to the normal rhythm of cell growth as it was before they became cancerous; 2. To visualize that once treatment is complete, only three of the cancer cells remain in the Petri dish; 3. No intention is formulated, Laskow lets the will of God operate through his hands; 4. To offer an unconditional love to cancer cells, without specific intention; 5. To visualize the dematerialization of these cells.

The most effective intention/strategy (39% inhibition) was to allow cells to go back to their normal growth. Giving way to the will of God allowed a 21% inhibition. Visualizing that there were only three cells left in the Petri dish resulted in 18% inhibition. Unconditional love without specific intent produced no effect. During subsequent tests with Laskow, Rein discovered that the combination of an intention of a return to the natural order of things, with the visualization that only three cells remained alive in the box, increased the rate of inhibition of cell growth to 40%. Associating mental images with intention seems to generate better results.[107]

For the last test, Glen Rein asked Leonard Laskow to focus on each of the five intentions/strategies described above while holding five vials of water in succession. These vials were then used to prepare the tissue culture medium necessary for the growth of the cancer cells. Again, water treated with the intention of a return to natural order gave the best results, allowing a 28% inhibition of cell growth. As if the water had "stored" the intentions expressed and communicated them to the culture medium as well as to the cancer cells.

## The Sense of Being Stared At

Some researchers, including British biologist Rupert Sheldrake, have conducted experiments to study the sense of being stared at.[108] This is a common phenomenon that occurs frequently in public places with strangers. This impression often leads people to turn around and notice that they were truly being observed.

One of the research protocols that was used by Sheldrake is direct observation. During the tests, the participants must quickly guess, in less than 10 seconds, whether they are being observed or not. Their answers, which are right or wrong, are recorded immediately; a session lasts less than 10 minutes and consists of 20 tests. These direct observation tests, which are easy to perform, were conducted with thousands of participants. Many of these tests were conducted in schools and universities. The results are very consistent and show approximately 55% of correct answers as opposed to 50% chance. This is a result that becomes extremely significant when repeated over the course of several thousand trials. In some of the direct-observation experiments, the participants were tested repeatedly, and the results were revealed to them after each trial. This training approach resulted in a significant improvement in the results.[109]

The other experimental protocol that was used by Sheldrake and other researchers involved separating the observer from the observed; they are placed in different rooms connected by a closed-circuit television system. The use of this protocol began in the 1980s. In these tests, participants are not asked to guess whether they are being observed or not. Instead, their skin conductance is recorded. These tests are composed of random series of tests, during which the observers either look at the image of the participant observed on the TV screen or look away thinking of something else.

Statistically, many of the studies conducted using this research protocol yielded significant results. These results show that skin conductance changes when observers look at the observed, even if the observed participants are not aware of it.[110]

Dean Radin has developed a variant of this experimental protocol to study the impression of being stared at. In this protocol, an electrogastrogram (or EGG) is used to check whether the observed's contractile activity of the gastrointestinal region, a zone particularly sensitive emotionally, can be influenced by the emotions of the observer who is in a different room, located at distance, and is watching a TV screen. EGG activity is measured in an individual who is in a state of relaxation. Meanwhile, in the other room, the other participant playing the role of the observer is watching the TV screen. The image of the person being observed appears as well as, periodically, emotionally positive, negative, neutral or calming video images.

The results showed that EGG responses were significantly greater when the observer experienced positive and negative states compared to at the emotionally neutral state. In other words, the observed's contractile activity of the gastrointestinal region responded to the observer's emotions, even though the pairs of participants (the observer and the observed) were physically separated and could not communicate with each other *via* normal sensory channels.[111]

The psychology researcher Stefan Schmidt, who runs the evaluation section of complementary medicine research at the university clinic of Freiburg in Breisgau, Germany, conducted a meta-analysis of about 30 studies on the non-local effect of intention on the skin conductance of distant human receptors or on the impression of being observed. This statistical analysis confirmed the existence of these phenomena.[112] The meta-analysis also revealed that the magnitude of these was correlated with the quality of the studies: the higher the quality level of the studies, the higher the magnitude of these phenomena.

## Distance Healing

In 1975, an English trainee in general medicine suddenly fell ill. She had symptoms suggestive of sepsis and meningitis. Dying, she was admitted to the hospital. The doctors then discovered that she was suffering from the Waterhouse-Friderichsen syndrome – an inflammation of the adrenal glands leading to acute adrenal insufficiency. This is caused by a severe bacterial infection that is associated with several risk factors. An antibiotic should be introduced at the earliest, together with intravenous therapy. If this syndrome is not treated as quickly as possible, it evolves very rapidly towards a coma and ultimately to the death of the patient.

The evening following the hospitalization of the young woman, individuals gathered and prayed for her to heal completely without residual disability. That night, at the same moment, the condition of the trainee suddenly improved. Doctors noted that the pneumonia that had infected the middle lobe of her left lung had now vanished. They were amazed.

The patient came out of the coma four days later. Upon her arrival at the hospital, an ophthalmologist discovered that there had been a hemorrhage in her left eye, which would cause permanent blindness to that eye. The young woman regained perfect sight of the left eye, as well as perfect health.[113]

This medical anecdote brings about the following question: can prayer for others really heal?

Psychiatrist and researcher Elisabeth Targ was interested in psi phenomena, as well as in the role of spirituality in healing and health and the therapeutic effect of prayer. In 1995, with AIDS still claiming many lives, she and her colleagues at the University of California Medical Center in San Francisco (UCSF) conducted a double-blind pilot study involving 20 advanced AIDS patients. All patients received conventional care, and spiritual healers prayed for 10 of them. These healers lived an average of 2,400 km from the patients. They did not know which group they were assigned to (randomly). As a result, they were unaware whether someone was praying for them. The study lasted six months. Four patients died, which represented a typical mortality rate. When the results were released, the researchers learned that the four patients who passed away belonged to the control group. In other words, the 10 members of the experimental group, for whom prayers were said, were still alive.

Encouraged by these results, Elisabeth Targ decided to conduct a confirmatory study the following year. Forty patients were recruited. They knew that by agreeing to participate in the study, they had a one in two chance to benefit from the prayers of the spiritual healers. Pictures of the 40 participants were taken. The participants were divided into the two groups (experimental and control) randomly, depending on the age, the level of activity of the immune system, and the number of complications related to AIDS. The photos of members of the experimental group were sent to 40 healers (for example, rabbis, shamans, and energy practitioners), while the photos of members of the control group were locked away in a drawer. The healers, who were far from the patients belonging to this group, performed their rituals for one hour a day for six consecutive days. Each week for 10 weeks, the healers switched patients so that each member of the experimental group could benefit from their prayers.

Where the patients were placed was revealed six months after the study started. The results revealed that patients in the experimental group had spent 600% fewer days in hospital and had contracted 300% fewer AIDS-related illnesses (e.g., ulcer and encephalitis) compared to patients from the control group. These differences between the two groups were significant. In addition, the placebo effect was not considerable as about 55% of the patients in both groups believed that the healers were praying for them, and they didn't feel any better than the other patients.[114]

The results of the two studies conducted by Elisabeth Targ and her colleagues are quite clear. However, this is not the case for all the studies on the therapeutic effects of prayer. Some of them suffer from methodologically significant shortcomings, and other studies produce results that are not as repeatable as lab-based studies. In all likelihood, this is due to the fact that many factors are involved in prayer for the healing of others, and it is difficult to control for all these variables. For example, the ability of participants to generate non-local intent effectively can vary considerably from study to study.

Hence, Dean Radin and his colleagues asked themselves how meditation training can modulate the effects of compassionate distance healing.[115] In their double-blind study (which they called the *Love Study*), they recruited couples. For each couple, one member was in perfect health while the other member was receiving cancer treatment. The healthy partner played the role of the sender, while the patient was the receiver. In the trained group, the senders were invited to develop a compassionate intent, an act of selfless love for another person to relieve their suffering and increase their well-being. The meditation training program consisted of a one-day workshop. For a period of three months, participants chosen to be the senders were required to practice meditation for 30 minutes daily.

In addition to the coaching group, there was a control group on the waiting list. The members of this group were the same couples but tested before the health partners followed the training program. A third group, the control group, consisted of healthy couples who did not receive training. When a couple arrived at the laboratory, they were both equipped with electrodes to monitor skin conductance, heart rate, and respiration. The two spouses were then asked to maintain a feeling of connection uniting them for the duration of the experiment. In the control group, the couple were asked to determine which of the two spouses was the most receptive – the role of receiver was attributed to this person. In the other two groups (trained and waitlisted), the receivers were still the cancer patients.

The receivers sat in a reclining chair in a room shielded from electromagnetic waves and acoustically isolated. They were informed that they would be seen at random time periods, *via* a video monitor, by the senders who were in another room (the two rooms were separated by a distance of 65 feet). During these periods, the sender had to make a mental effort to connect with the receiver. At other times, the video image disappeared, and the sender had to relax and not send any compassionate thought. The receivers did not know the length of the periods of compassionate intention or relaxation. The intentional periods, in fact, lasted 10 seconds, while the relaxation periods varied from 5 to 45 seconds. This duration was determined randomly by a computer program.

In all, 36 couples participated in the study. The analysis of skin conductance in all senders of the different groups revealed a significant increase after the images of the receivers had appeared on the video monitor. This indicated activation of the sympathetic nervous system resulting from the mental effort associated with compassionate intention and healing thoughts. About two seconds after the video image appeared, the skin conductance of the senders began to increase and then peaked around three seconds later. Radin and his colleagues also compared the skin conductance of the receptors in the three groups. All receivers responded promptly after the sending period started. In the control group, the skin conductance response disappeared after four seconds, whereas this response faded after five seconds in the control group on the waiting list. For the trained group, the skin conductance of the receptors increased gradually for eight seconds. It was in this group that the highest increases were noted. These results suggest that meditation training can increase the effects of compassionate remote healing.[116]

## Chapter 5

# A Delocalized Consciousness

*The day which we fear as our last is but the birthday of eternity.*
Seneca

## Near-Death Experiences and their Various Elements

What happens after the death of our physical body? Do our minds and personalities survive? What about our consciences? Do they vanish completely into oblivion? If life continues after death, do we continue to exist as an individual entity or rather within a universal Spirit? These deep existential questions have probably haunted human beings since the appearance of *Homo sapiens*.

There are multiple ways to tackle these major questions. One can, for instance, simply accept the teachings on this subject from religions, philosophers or even esoteric traditions. However, it is also possible to consider what science has discovered about death.

Over the last 40 years, several researchers have been studying the long-lasting mental experiences of individuals who are physiologically or psychologically close to death. These have been called "near-death experiences" [NDEs] and individuals who report such experiences are referred to as "experiencers." NDE can occur as a result of cardiac arrest, an accident, illness, or complications associated with surgery or severe bleeding. These experiences can also occur during or after childbirth as well as following coma, intoxication or electrocution.

Statistical studies indicate that in the Western world about 4% of the population has had an NDE. Other studies reveal that over the last 50 years, more than 25 million people worldwide have experienced an NDE.[117] This is already major, but the development of automatic defibrillators and access to this technology should significantly increase the number of experiencers in the future.

Apparently, NDEs are not influenced by factors such as gender, race, socioeconomic status, level of education, and belief systems. Thus, NDEs occur as frequently among atheists and agnostics as among religious people.[118]

Even though no two NDEs are identical, and experiencers hardly ever experience all the typical aspects of this experience, research shows that the NDE is characterized by some basic elements.[119] One of the most common elements is the out-of-body experience (OBE), which is the feeling of leaving one's body and seeing events unfold from an outside point of view or a spatially distant place. During OBEs, experiencers realize not only that they retain their mental faculties (e.g., consciousness, perception, thought, memory, personal identity, will, and emotions) but that these faculties are strengthened. Thus, hearing is sharp, vision can be extended to 360° and the mind is alert. Experiencers also frequently state that in the absence of their physical body, they are able to project to wherever they wish (e.g., pass through walls) and read thoughts.

From a scientific viewpoint, OBE is crucial as it is the most objective component of NDE, and the only one that can be corroborated independently. All other components of NDEs are essentially subjective. Feelings of peace and joy and passing through a dark region or dark tunnel that is often followed by the appearance of an unusually bright light are other typical aspects of NDE. On the other hand, it is not uncommon for experiencers to allude to another world of splendid beauty with striking cities and prodigious gardens, and also hear extraordinarily ethereal music. Experiencers also often report meetings with deceased family members or friends, who appear younger and healthier than in their memory.

In addition, experiencers commonly claim to have met a

"Being of Light" who can communicate telepathically, and who radiates total acceptance and unconditional love. Some cross-cultural studies indicate that personal religious beliefs and cultural perspectives influence how NDE is interpreted.[120] For example, a Christian might believe that the "Being of Light" encountered was Jesus, while a Buddhist might think that it was Buddha. In the presence of such a being, experiencers sometimes experience a review of life. During this review they can see their entire life unfolding very fast before them, and relive major or minor events in this life, sometimes by adopting the perspective of the other individuals involved. This element of the NDE allows experiencers to draw conclusions about their lives and the changes they may be subject to.

In the later phase of NDE, experiencers may encounter a barrier, in the form of a wall, bridge or river. There, they understand that they will no longer be able to return to their bodies and resume their lives if they cross this border. The "Being of Light" or departed relatives then explain to experiencers that they must return to their bodies because the objectives of their physical life have yet to be achieved. In general, it is at this point that experiencers are compelled to return to their bodies.[121]

## A Life-Altering Experience

Research on the impact of NDE reveals that these experiences positively and significantly transform the majority of experiencers. They share a comparable psychological profile after NDE, regardless of differences in personality prior to this experience: this is reflected in their beliefs and values, their attitudes and behaviors, as well as their visions of the world. These psychological and spiritual changes, which become more apparent over time, do not correspond at all to the kind of changes one would expect if one takes NDE as a simple hallucination.[122]

Let's take a brief look at the various types of changes

frequently reported by experiencers. First, when they come back, the way they perceive themselves has changed – they have better self-esteem and more self-confidence, reducing their reliance on the approval of others. They also realize that the main purpose of their existence is to discover and fulfill their mission on Earth. Experiencers marvel at and are filled with gratitude for life, which has more meaning and interest and for which they now feel a sense of deference. As a result, their concerns now extend to all life forms, and they become more sensitive to the health of the biosphere. Moreover, less critical and more tolerant, experiencers also become more altruistic and compassionate, which often leads them to donate to charity or to focus on humanitarian causes. They also spend more time with loved ones and friends, and they consider the quest for power, fame, and material success as pointless and empty.

Experiencers also often note a decline in their religious affiliations, which is accompanied by an increase in their spirituality and a greater interest in contemplative practices such as meditation and retreats. Moreover, irrespective of their previous beliefs, experiencers now have the certainty that life continues after the death of the physical body and that God (or a transcendent principle) exists. Experiencers also often return from their experience with various psi abilities (for example, clairvoyance, telepathy, and precognition), and they sometimes try to recover knowledge acquired during NDE and lost during "reintegration" into the physical body.[123]

Even though NDEs are powerfully transformative experiences, they not only have positive consequences, as most experiencers come out shaken, struggling to accept and integrate them. This process is even more difficult when their experiences are viewed negatively or skeptically by caregivers, family and friends. It is not unusual for experiencers to feel depressed, lonely and nostalgic during the first years after their experience.[124]

In a small number of cases, reported NDEs are unpleasant or downright scary. They may, for example, be accompanied by anxiety due to the experience of a loss of control during passage through the tunnel or "absolute void." The few studies available on negative NDEs suggest that egocentric people and those who are afraid of dying, who feel anger or guilt, or who believe they will go to hell when they die are more likely to experience a negative NDE.[125]

On this subject, here is an extract from the testimony of an individual reported by my good friend, Dr. Jean-Jacques Charbonier, a French anesthesiologist who has been conducting research on NDE for several years.[126] The negative experience reported took place during heart failure.

*There were a lot of people around me trying to revive me. I then went into a sort of very dark cone that was spiraling and a very strong current carried me to the end of this funnel. Along the way, I came across wrinkled, grimacing faces that seemed to belong to people who were suffering horribly. The more I sank into that cylinder that was getting narrower, the more the people were suffering. They were screaming but no sound came out of their mouths. It was horrible. I too could feel their suffering [...] When I finally arrived in front of this immense flame, I first thought that I was in hell and that I was going to be burned right away. But the flame started to dance funnily, and it enveloped me, asking me how I had helped others. I did not know what to answer. It was then that I realized that my past life had been a series of petty larcenies and scams.*

However, most negative NDEs also contain positive elements which often make up most of the experience. This type of NDE can also have positive consequences since most experiencers who have had a negative NDE are no longer afraid of dying, and they recover their faith in God. In addition, they interpret their negative experience as an opportunity to change their lives completely.[127]

## Confirmed Perceptions during Clinical Death

Fifteen years ago, neurologist Olaf Blanke and his colleagues reported in the scientific journal *Nature* the strange experience undergone by an epileptic woman.[128] Since the patient's seizures could not be treated by medication alone, she was advised to turn to neurosurgery. In order to obtain information on the location and extent of the epileptogenic zone (the area of the brain that causes the seizures) that had to be removed during the operation, the medical researchers implanted electrodes in her right temporal lobe. Other electrodes were implanted to identify and locate, through electrical stimulation, brain regions that, if removed, would cause paralysis or deficiency in sensory or linguistic abilities. Such an approach preserves the integrity of important brain areas that are adjacent to the epileptogenic zone.

When the researchers stimulated the angular gyrus – a brain structure located at around the temporoparietal junction, which seems to be involved in the perception of our body in space, the patient described herself as "floating" near the ceiling. She also claimed to see her legs "shrinking." In another article published a few years later, Blanke and his colleagues described the OBE as the outcome of a temporary dysfunction of the angular gyrus.[129]

This interpretation is not consistent because the experience reported by this epileptic patient represents an illusion involving a false perception of reality. Whereas, in contrast, typical OBEs associated with NDEs often involve verifiable perceptions of events related to resuscitation or accident, as well as surrounding areas. In this regard, some accounts from experiencers have been supported independently by witnesses. One of the most famous cases of veridical perceptions corroborated during NDE – perceptions that we can attest to have coincided with reality – is that reported by cardiologist Pim van Lommel and his colleagues in the Netherlands. In this particular case, the veridical perception was corroborated by a nurse in a cardiology intensive care unit during the pilot phase of a study conducted by van Lommel and his colleagues.[130] Here is the testimony of this nurse:

*During a night shift, an ambulance brought a 44-year-old man, cyanotic and in a coma, to the coronary care unit. Passers-by had discovered him an hour earlier in a field. After admission, he was given artificial respiration without intubation, while cardiac massage and defibrillation were performed. When we tried to intubate the patient, we had to remove the denture he had in his mouth. I removed these upper dentures and placed them on the "emergency cart." Meanwhile, the cardiac resuscitation attempt was ongoing. After about an hour and a half, the patient had sufficient heart rate and blood pressure, but he was still comatose, intubated and on a ventilator. He was transferred to the intensive care unit so that artificial respiration could continue. It was only after more than a week that I met the patient again, who was back at the cardiology division. While I was handing him his medicine, he called out to me:*

*"Oh, this nurse knows where my dentures are."*

*I was stunned. Then he explained to me:*

*"Yes, you were there when I was brought to the hospital and you took my dentures and placed it on the cart with all the bottles and sliding drawer underneath where you put it."*

*I was amazed because I seem to recall that scene unfolding while this man was in a coma and in the process of being resuscitated. I asked him to tell me more. The man told me that he had seen himself levitating over the bed, with the entire medical team busy around his body. He was also able to describe correctly and in detail the small room where he had been revived as well as the people who were with me. While observing the scene, he was fearful that we would cease the resuscitation protocol and let him die. And we did have a very negative prognosis for this patient considering his condition when he arrived. The patient also told me that he had tried desperately, but in vain, to make us all aware that he was still alive and that we should continue the resuscitation. He had been deeply affected by his experience and said he was no longer afraid of death. Four weeks later, he left the hospital in perfect health.*

Another interesting case of confirmed perception has been reported by the English cardiologist Richard Mansfield, who revealed the following to Dr. Sam Parnia.[131] During his night shift, Mansfield was urgently called to attend to a cardiac arrest. Along with the other members of the medical team, he rushed to the patient, a 32-year-old man who had no pulse, was not breathing and whose electrocardiogram (ECG) had flatlined. Dr. Mansfield and his colleagues then attempted to resurrect this man, although the chances of saving him seemed very slim. The patient was intubated and administered oxygen, adrenaline and a cardiac massage. Despite this intervention, the ECG remained flat and the pulse absent. Members of the medical team continued their efforts for over half an hour, unsuccessfully, then Mansfield told them to stop everything: it was terrible, but they had to accept the fact that the young man had died. Just before ceasing cardiopulmonary resuscitation maneuvers, Richard Mansfield looked at the monitor, the electrodes, and the wires attached to the electrodes to make sure everything worked perfectly.

Mansfield left the room and went to the nurses' station to write notes in the medical file of the patient who had just died. While carrying out his task, Dr. Mansfield realized that he did not remember exactly how many adrenaline vials had been administered to the deceased patient. About 15 minutes later, he returned to the room to check this detail.

In the bedroom, Mansfield looked at the young man and

found that his lips and skin were no longer as bluish as when the resuscitation maneuvers had ceased. Suspecting that something was wrong, Richard Mansfield approached the patient to see if he could feel a pulse in his groin. Stunned, the cardiologist discovered that the patient did have a pulse. The other members of the medical team were called on the spot and the cardiopulmonary resuscitation maneuvers were resumed. This time they were successful. Once the patient was stable, he was transferred to a room located in the intensive care unit.

About a week later, Mansfield visited the patient. To the cardiologist's amazement, not only had the young man fully recovered, but he had also suffered no brain damage. The patient then told the cardiologist that he had witnessed everything that Mansfield had done (check his pulse, stop the resuscitation maneuvers and leave the room, return later and once more check his pulse, and then restart the resuscitation maneuvers).

Richard Mansfield confided to Sam Parnia that he was confused by this case because it was a medical impossibility.[132]

Such cases are of great importance because during cardiac arrest, the flow of blood into the brain is interrupted and breathing stops. When this happens, the electrical activity of the brain (as measured by the EEG) disappears after 10 to 20 seconds, and the cardiac arrest victim is then considered to be in a state of clinical death.[133] In such circumstances, Dr. Jean-Jacques Charbonier proposes the use of the term "Provisional Death" since the cardiac arrest victim is no longer in "a state of imminent death": in fact, the patient has sometimes been clinically dead for several minutes.[134]

Because brain regions mediating mental functions are no longer functional, cardiac arrest victims are not expected to have clear conscious experiences that they could recall. However, studies conducted in the Netherlands,[135] Belgium,[136] the UK[137] and in the United States[138] have revealed that about 15% of cardiac arrest survivors recount memories of the time they were clinically dead. These studies, which evidenced more than a hundred NDEs, demonstrate that consciousness, perceptions, thoughts and emotions can be experienced during a period when brain activity is no longer detected by an EEG.

Proponents of materialistic theories of the mind argue that even if the EEG pattern is flat following cardiac arrest, there may be residual brain activity in the deep structures of the brain, which would not be detected because of the limitations of this neuroelectric activity recording technique. This argument is valid since the surface EEG – measured using electrodes applied on the scalp – primarily records the activity of cortical neural populations.[139] However, brain activity recognized by neuroscientists as the necessary condition for conscious activity and mental functions – perfectly measured by the surface EEG – is clearly absent during cardiac arrest.[140]

## Biological Explanation Attempts That Fail

Materialist scientists have proposed biological explanations to explain the different aspects of NDEs. In my view, these attempts at explanation all fail to account for the richness and multidimensionality of these experiences, as well as their implications.

One of these hypotheses proposed by psychologist Susan Blackmore is that of the "dying brain"[141]: A lack of oxygen (or anoxia) during the death process could produce abnormal excitement of neurons in the brain areas associated with vision, and such abnormal excitement would result in the illusion of seeing a bright light at the end of a dark tunnel. However, this hypothesis contains several flaws. First, tunnel observations are absent from several stories of experiencers. Secondly, if anoxia played an important role in the production of NDE, most individuals experiencing cardiac arrest should report an NDE, but this is not at all the case.[142] On the other hand, when their oxygen levels drop significantly, people with poorly functioning hearts or lungs experience an "acutely confusing state," which they remember little or nothing of, and in which they are very agitated and confused. The contrast is striking with those NDEs experienced with lucid awareness, clear reasoning, and well-structured thought processes. Experiencers also remember the NDE precisely, completely, and permanently. Had Blackmore's hypothesis been correct, the illusion of seeing a tunnel and a light should be stronger as the level of oxygen in the blood drops. However, individuals with low oxygen levels do not witness a tunnel or light, or other typical aspects of NDEs, and in some cases, some people experience NDEs even though they are not on their deathbed and oxygen levels in their brain are normal.[143]

Another research field indicates that the anoxia hypothesis is not valid. It has to do with the work done during the 1990s by Dr. James Whinnery of West Texas A&M. He conducted studies simulating the extreme conditions that occur during air combat maneuvers. During these studies, Whinnery subjected fighter pilots to significant gravitational forces in a giant centrifuge. Rapid acceleration significantly reduced blood flow and the delivery of oxygen to the brain. This caused the pilots to have short periods of unconsciousness, which Whinnery called "*dreamlets.*" The main characteristics of these *dreamlets* were confusion, impaired memory for events just before the loss of consciousness, and disorientation upon waking. These symptoms do not typically accompany NDEs, and no psychospiritual transformation has been reported following dreamlets.[144]

It has also been suggested that NDEs do not occur during the time of brain damage but just before or after, while the brain is more or less functional.[145] The weakness of this interpretation lies in the fact that the state of unconsciousness induced by the stopping of the heart leaves amnesic individuals confused as to the events that took place immediately before or after the cardiac arrest.[146] In addition, the confusion accompanying the loss or regaining of consciousness does not transform life at all, unlike NDE. On the other hand, in some cases, these experiences contain time markers that take the form of truthful reports of events that occurred at the precise moment of brain damage.[147]

Other researchers have speculated that hypercarbia, an increased level of carbon dioxide in the blood, could be involved in NDEs. However, the different aspects of NDEs are rarely present in hypercarbia, and these experiences can occur while the level of carbon dioxide in the blood remains normal.

If anoxia and hypercarbia cannot be held responsible for NDEs, could these experiences be hallucinations that simply result from chemical imbalance in the brain?

It is well known that the ingestion of ketamine, an anesthetic agent that is sometimes used recreationally, can generate hallucinations. In small doses, ketamine can also lead to OBEs. This drug works primarily by blocking N-methyl-D-aspartic acid (NMDA) receptors. These receptors open normally in response to binding with glutamate, the most abundant excitatory chemical messenger in the human brain. Psychiatrist Karl Jansen hypothesized that inhibition of NMDA receptors could lead to an NDE.[148] However, experiences with ketamine produce strange images and are often frightening. Moreover, most ketamine users know that the experiences produced with this drug are illusory. In contrast, experiencers are absolutely convinced that they have experienced something real. In addition, most of the central elements of NDEs are not reported by ketamine users.

Some have speculated that temporal lobe epilepsy may induce all the typical aspects of NDEs.[149] However, an exhaustive review of the medical literature related to this neurological problem reveals that the main characteristics of the NDE do not accompany epileptic seizures located in the temporal lobes. In fact, the symptoms associated with this type of seizure include hallucinations, mental confusion, and negative emotional states.[150] And even though neuroscientist Michael Persinger has claimed that he can produce NDE using transcranial magnetic stimulation of the temporal lobes,[151] the experiences reported by study participants have little in common with the characteristics of NDEs.

## Experiences on the Threshold of Death

Experiences on the threshold of death are another phenomenon that suggests that life does not cease when the physical body stops functioning. Most medical staff who work in palliative care are familiar with these experiences. These are manifested in the days, hours, and even minutes that precede death. These experiences have been reported since time immemorial among various cultures.[152]

During threshold of death experiences, experiencers are in a state of expanded consciousness – as if the veil separating life from death was being raised – and they see loved ones who have previously died or spiritual beings unknown to them. Sometimes dying people reach out to these "characters" of their visions, whom they perceive as messengers coming to assist them in their transition to the afterlife. At times dying people speak directly to these characters, or they report to the doctors and nurses caring for them what these characters tell them.[153] In some cases, dying people hear wonderful music or they describe places of great beauty belonging to another world. Similarly, just before dying, the great inventor Thomas Edison said: "It's very beautiful here." And Steve Jobs, founder and CEO of Apple, exclaimed: "Oh wow! Oh wow! Oh wow!"

These experiences reduce the physical and emotional distress of the dying, and bring them comfort, peace, and serenity. At the end of life, approximately 90% of individuals are in coma stages too deep to communicate. However, of the 10% who remain conscious before their death, more than half of them report such experiences.[154]

Many doctors believe that threshold of death experiences

are just hallucinations due to drugs or wishful thinking. This can sometimes be true. However, there are cases where the hallucination hypothesis is invalid because the dying person surprisingly recognizes a person who he thought was alive but who had actually died recently. Dr. Bruce Greyson, psychiatrist and researcher at the University of Virginia, compiled many of these cases in a study published in 2010.[155] In this regard, Greyson wrote: "Cases like this provide some of the most convincing evidence about the survival of consciousness after the death of the physical body."

Here are some of these cases:

*In a car accident, a woman sees her father who died an hour earlier*
An indigenous woman is dying after a car accident when a stranger stops to offer his help: she tells him: "If you go near the reserve, please tell my mother that I am fine. Not only fine, but also very happy because I'm already with my father." The stranger finds her mother and gives her the message. The mother tells him that about an hour before the accident, her daughter's father died of a heart attack.

*A boy sees his mother's first love*
On his deathbed, a 7-year-old boy with leukemia tells his mother that he went to heaven. The boy also tells her that he has met a man who has introduced himself as his mother's first love. Following a car accident, this man became disabled. The mother is convinced that she never spoke to her son about this former boyfriend. She later discovers that this man died the day her son had the vision.

*Shared Death Experiences*
Dr. Raymond Moody is the American physician who drew the attention of the general public to NDEs in the mid-1970s. His

subsequent research led him to discover a related phenomenon, the shared death experience (or SDE – Dr. Jean-Jacques Charbonier uses the term "death perceptions" to refer to it156). This phenomenon is documented in his book *Glimpses of Eternity: Sharing a Loved One's Passage From This Life to the Next.*[157] Moody spoke with hundreds of people who experienced a SDE. This experience is reported by a healthy individual who has an emotional connection with another individual experiencing the transition from life to death of his physical body. A SDE seems to rest on an empathic sharing of the experience of the dying person.

Based on his own work, Moody offers the model narrative of such an experience in this book:

*A woman named Jane is sitting with her husband, who is terminally ill with cancer, after thirty years together. He has lost consciousness and, according to the doctor who is treating him, his death is imminent. [...] As she looks at him, a white haze rises and vanishes in the air above him. [...] Suddenly, a white light with dancing particles envelops the room, Jane, who feels a bit dizzy, realizes that she has left her body and is floating near the ceiling of the room. She looks down at herself, sitting near the corpse of her husband, which feels odd as she can sense him not far from her. She turns her head and sees him smiling at her [...]. The couple continues to levitate with scenes from their lives unfolding around them. They travel through their past and witness these sometimes-panoramic fragments unfurling; [...]. Amongst these are scenes where Jane is not present, scenes from her husband's life. [...] Together, they move to a corner of the room that is no longer at right angles. The whole room's shape has changed shape and seems to keep on changing [...]. This may be due to this opening, this tube that seems to expand near the ceiling, like a door leading to another world. Jane and her husband go in [...] [and] find themselves in an Edenic landscape. There is nothing but splendor around them.*

*[...] Jane and her husband take a path that leads down to a stream. As they get closer, Jane realizes that she cannot go further. [...] She is happy for her husband who is not in pain and who does not inhabit his mortal body anymore. She takes leave of him and, in a flash, finds herself in her own body, sitting next to her husband's motionless corpse [...].*

The work of Raymond Moody indicates that several elements characterize SDEs:

*Perception of spiritual entities entering the room.* Experiencers see these apparitions, which seem to be the deceased loved ones of the dying, entering the room at the time of death.

*Vision of the spiritual essence of the dying leaving the physical body.* In some cases, experiencers report a mist (whitish or greyish) which emanates from the dead body and is sometimes a transparent replica of the person who has just died.

*Perception of a brilliant and mystical Light.* This supernatural Light, which illuminates the room, is perceived by the experiencers as being a source of purity, love, peace, and comfort.

*Participation in the life review of the deceased.* Experiencers can see scenes of the deceased person's life. These scenes sometimes form a hologram that surrounds the bed of the deceased individual.

*Impression of change in the structure of the room.* Experiencers sometimes report that when the other person dies, they seem to penetrate into another plane of reality that does not conform to the rules of three-dimensional geometry. For example, the cubic hospital room seems "to change" into something else; it can take the appearance of a funnel, and experiencers may have the impression of being outside their physical body and perceive the room from an angle that cannot be described with common words.

*Discovery of an unreal and Edenic landscape.* The experiencer discovers a landscape where all is splendid. They can also hear ethereal music so sublime that it is impossible to describe with words.

*Accompanying the deceased to the Light.* Some experiencers claim to have left their physical body at the moment where the individual they were accompanying died and reinstated their "bodily envelope" as the deceased entered the Light.

Like NDEs, SDEs can have a profound impact on those who live through these experiences, soothing the pain of death and reinforcing the belief in life after death. Furthermore, experiencers note an increase in their empathy, and they review their priorities leading them to make changes in their lifestyle. Feeling that they have suddenly matured and that their lives are gaining in depth, experiencers also begin to reflect on philosophical problems, and they no longer act impulsively. After their SDE, they also become more aware of their consciousness and of the importance of love for the other and the search for knowledge. Finally, experiencers often produce a calming effect on people.

As noted by Dr. Jean-Jacques Charbonier,[158] SDE cannot be secondary to phenomena of hallucinatory effects of anoxia or various cerebral disorders since the individuals who experience these are of sound body and mind.

## Can Light Have a Healing Power?

In late August 2014, Germaine Challut, who was then 79 years old, had severe abdominal pain. A few years earlier, she had been given an artificial anus because of rectal cancer. She was taken to the emergency room at the hospital. There, she was told that she was suffering from peritonitis with rupture of the intestine. The situation was dramatic because the stools had spread to her abdomen, and her body had become infected. Many of her organs were affected. As her condition deteriorated very quickly, a doctor decided to induce Germaine in an artificial coma, with the agreement of her daughter Isabelle, nurse and birth companion.

When the doctor decided to stop the artificial coma, Germaine did not wake up. After a week, Isabelle decided to "unplug" her mother's medical assistance because previous discussions with her revealed that Germaine was not in favor of being kept alive on a life support system. Her family came to the hospital to say goodbye. A priest was also present.

The next morning, to her amazement, Isabelle Challut found her mother sitting in bed, full of life and famished! Germaine told her daughter that she went through a tunnel moving towards the Light. She was no longer suffering pain, had no more fever, and a CT scan indicated that her abdominal infection had partially resorbed. Germaine agreed to take intravenous antibiotics for a week, and she was allowed to return home.

Her personality seemed also to have changed because while she had previously been usually an anxious woman, Germaine was now very calm and did not care about her physical condition. Back home, she gulped down broth and spent time connecting with the Light she had contacted during her NDE. Six weeks after Germaine's "return to life," her infection had completely disappeared, and she was doing very well.[159]

The case of Germaine Challut is not unique because other seemingly "miraculous" healings have been documented following an NDE. One of the most famous cases is Anita Moorjani, an Indian living in Hong Kong. In 2006, this woman, with stage IV lymphoma, became comatose and had an NDE. Following this experience, her cancer completely vanished. As with Germaine, this is completely inexplicable from a medical point of view. Obviously, such cases suggest that experiencers contacted a higher power that infinitely transcended the physical world and the domain of the living during their experiences.

## Chapter 6

# The Ultimate Mystery

*The beauty of death is presence, the inexpressible presence of
beloved souls, smiling at the tears in our eyes.*
Victor Hugo

## Communicating with the Deceased

In the previous chapter, we learned that the scientific study of
NDEs indicates that consciousness is not annihilated in the first
minutes after death. However, NDEs usually last only a few
minutes. They give us no clue as to what happens in the hours,
days, months or years following the death of our beloved ones.
This central question represents the ultimate mystery for many
human beings. To come up with an answer to this vital question,
it would be necessary to communicate with the deceased. But is
that possible?

This key issue has tormented visionary and open-minded
researchers for ages. In this respect, almost a century before
NDE studies began, Frederic Myers, Frank Podmore, and
Edmund Gurney – three pioneers of psychic research who had
contributed to the creation of the *Society for Psychical Research*
– gathered, in the 1880s, several thousand testimonies to study
after-death communications (or ADC) empirically.[160] ADC
occurs when the deceased spontaneously manifest themselves
to one of their relatives by a visual, olfactory, auditory or tactile
sign, without any help or external solicitation from a medium
or any other intermediary. Sensing or feeling the presence of
a deceased person is also part of ADC. Often, during these
experiences, deceased loved ones give assurance to the mourners
that they are now living happily in another plane of existence.

ADCs are a fairly widespread phenomenon, since studies

in some Western countries have revealed that about 30% of the individuals polled claimed to have had contact with a deceased.[161] These experiences make the grief of the bereaved more bearable and provide solace. ADCs also alter the beliefs of the bereaved about the meaning of life and death, and help them to cope with loss and mourning.

We cannot naively and simplistically assume, as do the pseudo-skeptics,[162] that all ADCs are only illusions or hallucinations because they are sometimes associated with objective phenomena, such as the movement or breaking of objects, and sounds that can be recorded. At times, a deceased person's apparition produces a shadow or is reflected in a mirror. Moreover, in some cases, more than one person may perceive the presence of the deceased, and in other cases, the deceased communicate new information to the bereaved which they can later check.

The case of James Chaffin's will is an obvious example. Arthur Berger, a lawyer from Miami, investigated this case in 1995.[163] Chaffin was a farmer from North Carolina who died in 1921. Sixteen years earlier, he drew up a will by which he bequeathed his farm to his eldest son, and nothing to his wife and to his other three sons. In 1919, James Chaffin decided to draw up another will. He placed it between the pages of a Bible and did not breathe a word of it to anyone. When the farmer died, his first will was the one to be validated. One night, after his death, one of his sons who had been left out of the will saw him at the foot of his bed. The apparition revealed to him that the new will was in the pocket of his old coat. After this appearance, the son found the coat and a note in his father's handwriting, sewn inside the pocket. Chaffin had indicated, on the note, the Bible reference where his son would find the latest will. The document was taken to James Chaffin's attorney. Although he found the story quite incredible, he could not question its veracity and the second will was validated.

Here are some recent ADCs listed by the After-Death Communication Research Foundation (ADCRF), an organization founded by Dr. Jeffrey Long, a doctor internationally known for his research on NDEs, and his wife Jody Long:

*Suddenly I woke up seated on the right side of the bed and saw my brother sitting on the wooden panel that served as a footboard (a head and a footboard). He looked just like I remembered him – dressed in jeans, and a leather jacket, with his black and wavy hair shining and his big black eyes sparkling. I do not remember being surprised at all; what was happening was normal. At the same time, I saw my husband sleeping on my left, as well as cardboard boxes containing clothes not yet put away, and also our disassembled wardrobe. My younger brother then spoke to me. He explained that he was not really gone, that he was doing very well, that he was happy and that I had to stop crying so much.*
*– Babette B.*[164]

*One night, I was asleep, and I know that this was no dream. I felt something, someone close against me and squeezing me tightly. I realized at that moment that it was him, it was his way of holding me at night, he was there against me. I remember taking his hand; I really felt it and squeezing it tight against my chest. I felt his breath on my neck and his distinctive smell. He was there, I'm certain. Since that moment, I feel much better even if I am still grieving and sometimes burst into tears; but he had come to reassure me. I know he is well. Since then, I have the impression that he is always there with me and that he accompanies me and protects me day and night. I still feel him at night next to me. He is not dead; he is alive, in another form.*
*– Marie M.*[165]

## Induced Post-Mortem Communication
ADCs seem to occur spontaneously, at least from the perspective

of the bereaved. But an American psychologist, Allan Botkin, claims to have developed a therapeutic technique that allows the bereaved to intentionally come into contact with their deceased relatives.[166]

In the 1970s and 1980s, Botkin tried to treat war veterans suffering from post-traumatic stress disorder (PTSD) by using so-called "confrontational" therapy. This classic therapy, prevalent at the time, consists in repeatedly confronting the patient with his memories of traumatic experiences in order to gradually weaken his emotional reactions. Botkin became frustrated by the inefficiency of this approach, which can take years to yield results. In 1990, he discovered EMDR (Eye Movement Desensitization and Reprocessing), another psychotherapeutic technique for the treatment of psychological trauma. Botkin realized that with EMDR, he obtained better results faster as he could see important changes sometimes within the first session.

During an EMDR session, the patient is asked to clear her mind and focus on a negative thought or image related to a distressing event. The patient must simultaneously follow the therapist's fingers which move laterally into the patient's field of vision for 20 or 30 seconds. The patient is then asked to focus on a positive belief, previously described at the beginning of the session, and to focus her attention once more on the painful memory. After several cycles of rapid eye movements, the patient's positive belief is anchored, and her emotional distress vanishes.

Five years after becoming familiar with EMDR, Botkin met a patient named Sam. He fought in Vietnam where he befriended Le, a young orphan whom he had promised to himself he would adopt and bring back home with him. One day, while Sam and other US Soldiers were helping a group of orphans, including Le, to get on a truck to take them to an orphanage, they were bombed. Sam then discovered Le's body lying in the mud. He was absolutely crushed until he met Allan Botkin. During his

EMDR session, Sam saw Le, who had now grown into a gorgeous woman with long black hair. He also felt her arms around him. Le thanked him for taking care of her before she died. Sam was jubilant, convinced that he had had genuine contact with Le. As for Botkin, he remained skeptical and believed that Sam had confused reality with the imaginary. Nevertheless, many patients react in a similar fashion to Sam, and these "contacts" seem to be soothing and reassuring, and also lead to a better acceptance of death. Botkin decided to follow this lead.[167]

His first Induced After-Death Communication (IADC) session was with Gary, whose daughter Julie died at the age of 11. Julie's mental competence never surpassed that of a 6-month-old baby, due to a severe oxygen deficiency at birth. After a heart attack, Julie died in Gary's arms. Although not a believer, Gary agreed to try Botkin's new method. In this method, the therapist asks the patient to focus on grief. After the eye movement, and once the feeling of grief has lessened, the therapist asks the patient to keep her eyes closed. This allows the patient to enter a state of receptivity conducive to receiving any information that may come from the deceased.

When he opened his eyes at the end of the session, Gary was utterly dumbfounded. He then told Botkin: "I saw my daughter! She was playing happily in a garden filled with vibrant and sparkling colors. She seemed perfectly healthy, without the physical problems she had experienced during her life. She looked at me and I sensed her love." Julie also told Gary that she would always be with him. When he emerged from the session, he was ecstatic as he felt that he had reconnected with his daughter. A year later, he told Allan Botkin that he still felt in touch with Julie. His IADC session showed him that "people do not really die, they just change state and inhabit a different environment, an exquisite environment."[168]

Allan Botkin has conducted several thousand IADC sessions. The results are impressive because the method works in 70% of cases, with people of all faiths, including atheists and skeptics. In the state of consciousness induced by this technique, the patient receives information in the form of visions and voices heard telepathically, and he may also feel physical contact. The deceased are often full of love for the living, whom they reassure and encourage to console themselves. In some cases, the deceased tell the living that they are forgiven for certain things or they ask for forgiveness for actions on Earth that they now regret. In addition, sometimes the deceased communicates new information to the patient (for example, a warning about a future situation like a health issue[169]).

Although Botkin refuses to speculate on the true nature of the communications induced by his method, in order to leave the patient free to interpret his experiences without being influenced by the therapist's beliefs, he, however, denies the objections that the IADC are hallucinations: "The questionable proof is that the consultants describe them as completely different from any other experience. Technically, hallucinatory perceptions are independent of sensory perceptions, that is, they are all in the subject's head, unrelated to external reality. Hallucinations often involve very different negative emotions from one person to another and are related to important psychological disorders. While the IADC most often involves positive content, similar from one person to another, they are very beneficial psychologically."[170]

According to Dr. Olivier Chambon – a French psychiatrist and psychotherapist who uses the Botkin method – this approach is revolutionary because it radically changes the way we think about mourning therapy, which it greatly fast-tracks. Like Allan Botkin, Dr. Chambon does not think that IADCs can be reduced to hallucinations, defense mechanisms or unconscious desires because sometimes the messages of the dead can be the opposite of what patients expect. Olivier Chambon also claims that contemporary specialists in grief therapy now recognize that the old recommendations to forget the deceased – like not having internal conversations and not keeping things belonging to them – "were unsuitable and potentially harmful. On the contrary, it is necessary to maintain the relationship which has finally changed level: this relationship has become more subtle, but the bonds of love can grow deeper." Thanks to the Botkin method, "the relationship is maintained, it is not over, it is just of a different nature; it remains dynamic, interactive, and people realize that their deceased have not suffered." The relationship with death of those who consult is automatically altered: "When they contact the deceased, patients know they will find them. They are less afraid of death and of their own death."[171]

## When Science Takes an Interest in Psychics

Psychics are known to be able to communicate with the deceased and receive information about living conditions in the afterlife. As for clairvoyance, it refers to all faculties specific to psychics.

During the second half of the nineteenth century, clairvoyance became very popular and pioneers of psychic research – such as William James, Edmund Gurney, Frederic Myers and Charles Richet – saw it as an important phenomenon which psychology had to focus on. For these researchers – who did not believe that mediumship could be explained simply by chance, suggestion, hallucination or fraud – it was imperative to empirically study the possibility that psychics could deliver authentic messages from deceased individuals without going through normal sensory channels; because these researchers firmly believed that this type of study could potentially demonstrate the survival of consciousness and personality after death.

The investigations of the so-called mediums during this period revealed numerous frauds, and the practice of mediumship quickly lost its credibility. In particular, the illusionist Harry Houdini (whose real name was Ehrich Weisz), who was fascinated by the question of survival after death, was one of those who led a fierce hunt for charlatans and impostors of all kinds: fake mediums. The unmasking of these impostors, who used conjuring, made sure that the majority of scientists of the time decided to distance themselves from the study of mediumship.

It was not until nearly a century later that serious scientists dared to study this phenomenon rigorously and systematically. Two of these scientists were Julie Beischel, co-founder of the Windbridge Research Center,[172] and my colleague and friend Gary Schwartz, director of the Laboratory for Advances in Consciousness and Health at the University of Arizona.

When a medium first encounters a client he does not know, that medium may be able to provide the client with a certain amount of factual information that he claims to have from deceased persons. The essential question that arises at this time is obviously: "Where does this information come from?" While doing a postdoctoral fellowship in Gary Schwartz's lab, Julie Beischel focused on this issue. Working with her research director, Beischel developed a rigorously controlled protocol which would make it possible to verify the nature of the information that can be received by mediums when "conventional" opportunities to obtain this information are eliminated.[173]

There are various "conventional" ways to get information about someone you do not know. First there is fraud or deceit: a person who is not a medium can carry out searches beforehand (for example, on the Internet) on the client, or the deceased, to come up with accurate information. The protocol developed by Beischel and Schwartz eliminates this possibility since the medium has only the first name of the deceased person throughout the duration of the experiment. Another conventional explanation is "cold reading." This expression refers to a set of psychological techniques used by fake mediums to analyze the visual or auditory cues that they get from clients. These techniques allow fake mediums to develop a correct and complete reading simply by observing the client's reactions to the information presented to them. In the experiments conducted by Julie Beischel, the person who plays the role of the client (the subject) is not physically present in the same room as the medium. In addition, the person conducting the experiment (the experimenter) does not know anything about the subject and has no information about the deceased. Hence, the experimenter cannot transmit subtle or unconscious signals to the medium. Finally, another possible conventional explanation is that statements made by the medium are so general that they could apply to almost anybody. To eliminate this last possibility, in Beischel and Schwartz's protocol, the experimenter asks the medium to answer four specific questions about the deceased person: physical description, personality, hobbies or activities, and manner of death.

In experiments conducted by Beischel, each medium was conducting a reading about two dead people with only their names as information. During the reading, the experimenter asked them the four specific questions on each deceased. The readings were done over the phone, and the experimenter was unaware of anything except the first names of the deceased. The medium and the experimenter were not in the same city, and often not even in the same state. When both readings were completed, they were then transcribed. Then, any reference to a name was removed from these documents, which were then given to read to people whose relative was one of the two deceased, without them knowing which of the two readings corresponded to them. These people had to evaluate the concordance of each of the readings with the deceased by rating them. This approach eliminated the possibility of evaluator subjectivity.

The many experiments carried out by Julie Beischel and Gary Schwartz indicate that mediums can receive unexplained accurate and precise information on dead people they do not know anything about. How is this possible? Beischel and Schwartz have proposed that two hypotheses can account for their results: either mediums actually communicate with the deceased, or mediums can "read" the minds of clients who come to consult them through clairvoyance or telepathy. The second hypothesis implies that mediums would obtain information without really communicating with the deceased. However, given that the information provided by the medium was sometimes unknown to the person who was undergoing the experience as a client, but the veracity of which was later confirmed, Beischel and Schwartz therefore believe that their results support the idea of the survival of consciousness. According to them, an aspect of our personality would continue to exist after physical death in a form capable of communicating with a medium.[174]

## Instrumental Transcommunication

The expression "Instrumental Transcommunication" (or ITC) was originally put forward by the German physicist Ernst Senkowski. Now deceased, he had long been a professor in electrical engineering at Polytechnic School of Rheinland-Pfalz. ITC refers to a means of communication with the "invisible world" that relies on the use of technology (for example, radio or television), for receiving messages, rather than on a medium. This approach, which has grown more polished in recent decades, therefore complements mediumship. Similarly, ITC is used to bring solace to individuals who are experiencing psychological distress, following the loss of a loved one. Another objective of ITC is to validate the existence of life after death.

In its current version, ITC is usually either audio or video. Audio ITC usually consists of an audio recording in which a background noise is added to help the deceased alter it. With respect to video ITC, "static" on a television screen is filmed with a camera, and then the recording is analyzed to detect images that seem "out-of-the-ordinary."[175]

The idea of using technology to be able to communicate with the dead is far from new. In the 1870s, American inventor and industrialist Thomas Edison – who filed more than a thousand patents over the course of his long career, and to whom we owe several inventions, including the alkaline battery, the light bulb, and the phonograph – wanted to offer the proof of the survival of personality thanks to technical advances. Edison was convinced that it would be possible to develop a "necrophone," a device capable of receiving voices from the afterlife. Thomas Edison was not the only one to think so because most pioneers of radio and television – such as Oliver Lodge, William Crookes, John Logie Baird, Édouard Branly and Guglielmo Marconi – believed that it was only a matter of time before technological development could enable communication with the dead.[176]

Edison's necrophone project failed. But important developments, related to the audio ITC, occurred a few decades after the death of the famous inventor. Thus, in the early 1950s, two Italian Catholic priests, Father Ernetti and Father Gemelli, worked together on a research project on music. Ernetti was an internationally recognized physicist, while Gemelli, who was a medical doctor and psychologist, was president of the Pontifical Academy of Sciences. In 1952, when Ernetti and Gemelli were trying to record a Gregorian chant, one of the tape's wires was constantly breaking. Exasperated, Father Gemelli looked towards the sky and asked his dead father for help. Shortly after, the two men, stunned, heard the voice of Gemelli's father, recorded on the tape. The voice said: "Of course I'll help you. I'm always with you." Gemelli and Ernetti decided to repeat the experiment. This time, a very clear and humorous voice said: "But, Zucchini, [...] do you not know it's me?" Father Gemelli looked at the tape on the recorder. Nobody was aware of the nickname his father used to tease him with when he was still a boy. Gemelli thought at that moment that it was his father who was speaking to him through the tape recorder.[177] Later, the two priests visited Pope Pius XII in order to share their experience with him. Although Pius XII was not surprised beyond measure by what Fathers Gemelli and Ernetti reported to him, it was decided, at the Vatican, that the matter should not be spread any further.

A few years later, in 1959, Swede painter and filmmaker Friedrich Jürgenson also realized, fortuitously, that the voices of the deceased could be recorded. Jürgenson's accidental discovery occurred when he decided to record on tape bird songs around Stockholm for the soundtrack of one of his films. Back home, while listening to the recording, Jürgenson thought he heard, overlapping the songs of birds, a weak voice calling out to him. He assumed that it was radio interference, but there was no transmitter at the place where the recording had been made. Intrigued, Jürgenson continued to experiment at home and he managed to pick up other voices and short messages from his mother or deceased friends.[178] Friedrich Jürgenson was so impressed that he spent the rest of his life recording voices.

After his meeting with Jürgenson, who had published a book about the many messages he had received, Konstantin Raudive – a Latvian psychologist and literary man teaching psychology at Uppsala University – took over from the Swedish filmmaker. In the years that followed, Raudive made more than 70,000 recordings using the static noise of a radio.[179] In 1971, sound engineers working for Pye Records, a famous British music record label, invited Konstantin Raudive to participate in a controlled experiment. Raudive agreed and the experiment was carried out in an acoustically insulated room that blocked radio waves. During the experiment, Raudive was not allowed to touch the tape recorder: he was only allowed to speak into a microphone. The recording lasted 18 minutes, and the sound engineers heard no sound other than Raudive's voice. Once the recording was completed, the engineers rolled back the tape to listen to it. They were astonished as they could distinguish more than 200 different voices.[180]

The large-scale works of Konstantin Raudive allowed him to determine four characteristics of electronic voices: different rhythm from ordinary language, extreme brevity of the message (usually a single word or a very short sentence), non-respect of syntax and sometimes a mixture of different languages.

Marcello Bacci was another researcher who was very successful in the electronic voice phenomena field. For nearly sixty years, Bacci has volunteered his ITC method to serve mourners. This man, who did not take advantage of this service and did not seek any advertising, used an old tube radio – which was tuned to white noise in the shortwave band – to help mourners get in touch with deceased people who are dear to them. To complete his ITC experiments successfully, Bacci used the Direct Radio Voice Method (DRV) – he tried to obtain communications directly through the radio. The results were spectacular as voices often addressed listeners by name and answered their questions by providing many relevant details. Over time, several journalists and scientists have participated in these experiments, and have apparently been able to converse with the deceased, without any fraud being apparent.[181]

Sarah Estep, another pioneer of electronic voice phenomenon, founded in 1982 the American Association of Electronic Voice Phenomenon.[182] Inspired by the approach publicized by Jürgenson then Raudive, Estep has received several thousand messages with only a tape recorder. She separated all these messages into three categories: 1) *Class C:* this is the category that most messages belong to. Often difficult to interpret, they are low whispers and can only be heard with headphones; 2) *Class B:* these messages are stronger and sharper, and they can be heard frequently without headphones; 3) *Class A:* these messages are strong and distinct, and they do not require headphones.

Concerning the afterlife, a summary of the information present in the clearest messages received by Sarah Estep follows:

- Family relationships can continue, if desired. Hence, instructors enlighten those who want to communicate with the Earth. However, the frequent calls of those left behind on Earth make the dead weary.
- In the hereafter, the dead lead active and productive lives, and they can continue the same activities as on Earth (sciences, arts, etc.). Their job is best suited to their aptitudes and personality.
- The deceased have the opportunity to visit several places, in the afterlife, before choosing one of them. They live in houses.
- Doctors and nurses help those who need it.
- Some of the dead continue to worship God and may belong to a "church."[183]

In the mid-1980s, some ITC researchers began to receive pictures of deceased individuals on television screens. German Klaus Schreiber, who worked with Professor Ernst Senkowski, was the first of these researchers. His technique was to direct a video camera towards a television and route the signal from the camera to the TV to create an "optoelectronic" feedback loop. The result was a kind of "static" on the TV screen, from where the faces of the deceased or images of places presumed in the afterlife slowly emerged. Using this method, Klaus Schreiber clearly recorded the faces of his two dead wives, that of his daughter Karin, as well as those of physicist Albert Einstein and movie star Romy Schneider. In 1988, hardly a few years after starting his ITC video experiment, Schreiber died of a heart attack. Apparently, a few months later, Klaus Schreiber started to send, to his old friends and colleagues whom he left behind on Earth, pictures of himself and the house he seemed to inhabit with other deceased members of the Schreiber family.[184]

Klaus Schreiber is not the only ITC researcher to appear to his colleagues who are still (physically) alive. Indeed, Friedrich Jürgenson, Konstantin Raudive, and other deceased pioneers have come into contact with ITC researchers quite often through different means. For instance, American Mark Macy, who contributed to the establishment of the International Network for Instrumental Transcommunication,[185] claimed to have received no less than seven telephone calls from Konstantin Raudive (who passed away in 1974). The conversation, during one of these calls, lasted almost 15 minutes. During another communication, *via* a radio, Raudive revealed to Macy and other colleagues that since his death, he has continued to work actively on the development of ITC systems from the other side. He also tells them the following: ITC works only if the "vibrations" of those attending the sessions are in perfect harmony and their intentions are pure.[186]

Gary Schwartz, my colleague at the University of Arizona, mentioned in the previous section, has been researching mediumship for many years. His research has led him to believe that although some psychics can sometimes be very precise, they are not infallible. As a result, he initiated the development of a communication system with the "invisible world" that would be less likely to be influenced by human subjectivity. In this context, a few years ago, this researcher in psychology tested in laboratory the possibility of using an ultramodern silicon photomultiplier to detect and record low levels of photons potentially associated with the presence of deceased individuals. Schwartz carried out a few proof of concept experiments. These experiments were divided into blocks of five minutes, during which the deceased were invited to manifest themselves in a dark room. These blocks were compared to control blocks of the same duration during which the experimenter (Schwartz himself) did not issue any invitation. Other blocks served to measure the impact of the intention of the experimenter. During all the blocks, the photomultiplier counted the number of photons detected in the dark room. The results showed that the sum of photons detected, when the deceased were invited, was significantly higher than in the other two types of blocks. These results suggest that a silicon photomultiplier can be sensitive enough to develop another type of ITC system.[187]

For pseudo-skeptics, the voices and images obtained in ITC can only be hoaxes, or aberrant interpretations of natural phenomena such as, for example, artifacts related to the technological equipment used. Such irrational interpretations could also result from apophenia – an alteration of perception, which leads an individual to attribute a particular meaning to what is really only statistical noise – or pareidolia, a kind of optical illusion which consists of associating a formless and ambiguous visual stimulus with a clear and identifiable element. This can happen when the voices or images obtained are not distinct at all. But what about when the voices or images received are clear and unequivocal? In these cases, it is the interpretations of pseudo-skeptics that miss the mark. On the other hand, as noted by Dr. Jean-Jacques Charbonier,[188] several electroacoustic experts have meticulously analyzed the voices recorded in the laboratory using the most sophisticated means. These experts accept that the phenomenon of electronic voices cannot be explained given the current state of scientific knowledge. The analysis conducted by these experts determined that the voices perceived in ITC can exceed 1400 hertz. Such frequencies cannot be reached with human vocal cords, which produce voices between 80 hertz for the lowest frequencies and 400 Hz for the highest frequencies.[189]

In my opinion, the most plausible and respectful explanation of the ITC principle of Occam's razor[190] is that the consciousness and personality of the deceased continue after the death of the physical body. This does not mean that the images obtained in

ITC are exact replicas of deceased individuals or buildings in any plane of existence. I think rather that the voices and the images received are the product of a transduction of information,[191] which is carried by the energy associated with the mental activity of the deceased. If this is correct, the recorded voices or images would be the last relay in this informational transduction process. In the last chapter of this book, I will come back to this process and the subject of energy associated with the activity of the psyche.

## Chapter 7

# Past Lives or Fantasy?

*Even chance meetings are the result of karma... Things in life are fated by our previous lives.*
Haruki Murakami

Jürgen Keil is a psychologist researching reincarnation – the process of survival after death whereby an immaterial and individual principle (for instance, the "soul") would complete one life passage after the other in different physical bodies. In 1997, he studied the case of a 6-year-old Turkish boy named Kemal Atasoy who claimed he had excellent recollection of a previous life. Kemal lived in a comfortable house located in an upper middle-class neighborhood with his well-educated parents. Since he was two, he claimed to have lived in Istanbul, 800 km from the family home. According to Kemal, in that past life, his family name was Karakas, and he had been a wealthy Armenian Christian living in a large three-story house by the waterside. There were moored boats and a church behind the residence, which was located next to that of a woman named "Aysegul." Karakas only lived there for part of the year. He often carried a large leather bag, and his wife had a Greek name.

Kemal's parents were neither acquainted with any Armenian nor anyone in Istanbul. And they were skeptical about Kemal's statements. Despite this, Jürgen Keil decided to verify whether the child's claims could be confirmed.

With his interpreter, Keil went to Istanbul. There, he found Aysegul's house, with an empty three-story residence near it, perfectly matching Kemal's description. This residence was found near the water, and there were indeed moored boats and a church behind. Jürgen Keil also spoke with an old man in the neighborhood, who told him that an Armenian had actually lived in this residence. The following year, Keil returned to Istanbul and met with a well-respected local historian. The historian confirmed that a wealthy Armenian Christian had lived in that house. His last name was "Karakas." He was a merchant and often carried a large leather bag. In addition, his wife was Greek Orthodox, and they lived in the house only during summer. He died in the early 1940s.

How could Kemal Atasoy, who lived in a small town 800 km from Istanbul, know so much about the life of a man who had died fifty years before his birth? For this little boy, the explanation was simply that he had been Karakas in a past life.[192]

## Research on Reincarnation

Through the ages, many peoples such as the Ancient Greeks, the West Africans, and many Northwestern Amerindians in North America, as well as followers of religions like the Hindus, Buddhists and some Christian groups have maintained their belief in reincarnation. This belief is still present as surveys have revealed that nearly 25% of Americans and Europeans believe in reincarnation.

Dr. Ian Stevenson, a psychiatrist who was for a time Director of the Department of Psychiatry at the University of Virginia, spearheaded research in this sector. Stevenson began his work during the 1960s. In the course of his career as a researcher, he focused on cases from India, Sri Lanka, Thailand, Myanmar (Burma), Lebanon, Turkey, Nigeria, Brazil, and the United States. During the following decades, other researchers – such as Erlendur Haraldsson, psychologist at the University of Iceland, and Jim B. Tucker, psychiatrist at the University of Virginia – started to do research on reincarnation.

In 1967, Dr. Stevenson founded the Division of Perceptual Studies at the University of Virginia. Jim Tucker is now the head of this research group. During the last fifty years, members of this group have identified more than 2,500 cases of children claiming to remember events that took place in previous lives.

Researchers in this area find cases more easily in countries where belief in reincarnation is widespread. They find such cases mainly through newspaper articles and by word of mouth. Researchers usually travel to the site once the child has given enough details about the past life to allow his current family to contact his previous family.

The investigation begins with an interview with the child's current family. The parents first provide information to the researchers about the child's statements and behavior. The researchers then verify the similarities between the child's assertions and the life of the deceased individual (the previous person). When a child has a birthmark that appears to match a wound on the body of the deceased, the researchers try to determine the degree of similarity with the help of the autopsy report – whether it is a lethal injury, or the medical record.

Children's claims are the main subject of the cases examined by researchers. Children make assertions when they are very young. Most children make these claims when they are between two and four years old and usually stop talking about their previous life around the time they turn six or seven. Many of these children exhibit behaviors and emotions that seem to be associated with memories of the past lives that they report. For example, sometimes children cry and beg their parents to take them to their former family. In addition, many children develop phobias that seem directly connected to the way the previous person died. Furthermore, some children are constantly replaying the scene of the previous person's death, and when that person has been murdered, they may feel a lot of anger towards the murderers. In other respects, certain children prefer. the favorite foods of the previous person, even tobacco or alcohol. As well, oftentimes when the child's family is driving them to where the previous person lived, they seem to recognize members of the deceased person's family as well as places related to their previous life. On the other hand, several individuals who were subject to studies by Stevenson and Tucker were born with birthmarks or congenital anomalies matching the usually fatal injuries on the previous person's body.[193]

In some cases, children claim to have remained close to the body of the previous person and the family, and to have attended the funeral. Thus, at two, Ratana Wongsombat, who was born in Bangkok in 1964, told her father that she had been a Chinese woman, named Kim Lan, in her past life. Ratana also told her father that Kim Lan's ashes had been scattered under the Bodhi tree of the temple, instead of being buried as she had wished. Kim Lan's daughter, Mae Chan, confirmed Ratana Wongsombat's statement, explaining that when she had tried to honor her mother's wish, the tree's roots had been so abundant that she ended up scattering her mother's ashes instead of burying them.

In other cases, children report experiences during the interval between death and reincarnation. For example, a boy named Lee recalled that he had decided to be reborn at some point. He also claimed to have been assisted by other beings in his decision to reincarnate on Earth. Similarly, Patrick Christenson, a little boy who had three birth defects matching those of his late brother Kevin, reported talking with a relative named "Billy the Pirate" in "Heaven" who told him that he had been shot and died in the mountains. Patrick's mother said she had never heard of a family member with that name, but when she called her own mother to inquire, she discovered that a cousin who bore the nickname "Billy the Pirate" actually died that way.[194]

## Birthmarks and Congenital Anomalies

Patrick Christenson, whom we just discussed, was born in Michigan in 1991. When the hospital nurses brought him to his mother, she immediately felt that he was connected to her first son Kevin. The latter had died of cancer at the age of 2, twelve years previously. Patrick's mother noticed that Patrick had three congenital anomalies that matched three marks on Kevin's body who began to hobble at 18 months old. One day he fell and broke his left leg. The doctors then gave him a checkup that included a biopsy of a nodule on the scalp, above his right ear. A bone scan revealed a cancer accompanied by numerous metastases. In addition, his left eye was bulging and bruised due to a tumor.

Kevin underwent chemotherapy, with a catheter inserted into a vein on the right side of his neck. After this treatment, he was able to go home. However, five months later, Kevin was hospitalized again. By then, he was almost blind in the left eye. He was treated with antibiotics because he had a fever. He died two days later, three weeks after his second birthday.

Shortly before his death, his parents separated. After that, Kevin's mother remarried, and she gave birth to a little girl and a little boy. Then she gave birth to Patrick. Patrick had a sloping birth spot, akin to a small cut, at the exact spot where the catheter had been inserted under Kevin's skin. In addition, Patrick had a nodule on his scalp, exactly where Kevin had his. Patrick also had a cataract in his left eye, causing almost complete blindness in that eye. This problem was due to corneal leukemia.

At about four and a half, Patrick started to make statements that seemed to be related to Kevin's life. For instance, he told his mother that he wanted to go back to their old house, and that he had left her there. He also claimed that this house was orange and brown, which was correct. Patrick asked his mother if she remembered the operation he had undergone. When she told him that he had not had one, he showed her the place above his right ear where Kevin had a nodule biopsy. Another time, Patrick saw one of Kevin's photographs, which were not usually displayed in the family home, and said it was a photo of him.

Patrick's mother learned about the research conducted by Ian Stevenson and Jim Tucker, and contacted them. The two researchers visited the family when Patrick was five years old. They obtained Kevin's medical records, as well as copies of Patrick's eye exams. Stevenson and Tucker noticed the opacity in Patrick's left eye. They also photographed the birthmark on his neck. The spot consisted of a dark, sloping 4-millimeter line on the lower right side of Patrick's neck and looked like a cut which had healed. The nodule on his head was difficult to see, but easily palpable.

The story of Patrick Christenson is one of the cases of birthmarks and congenital anomalies described by Ian Stevenson in his book *Reincarnation and Biology*.[195]

In another book, *Children Who Have Lived Before: Reincarnation Today*, German writer Trutz Hardo tells the story of a three-year-old boy from the Golan Heights region, near the border between Syria and Israel. It was Dr. Eli Lasch who was instrumental in the development of Gaza's medical system, who brought this fascinating story to Hardo's attention. The boy in question belonged to the Druze ethnic group. This religious community believes in the reincarnation of souls. The boy was born with a long red birthmark on his head. When he was old enough to talk, he told his family that he had died from a blow to the head, administered by a man with an axe. Since he knew the village from where he came, his parents took him there.

The boy also remembered the name of the previous person. A villager told the family that this person had disappeared four years earlier. The child then revealed his murderer's full name, and he also managed to find his former house. While walking with his family, the boy recognized the said murderer and called him by his name. Then he confronted this man saying: "I was your neighbor. We fought, and you killed me with an axe." All the blood drained from the man's face, but he did not confess to the crime. The boy then told the village elders where the body of the previous person was buried. At this exact same spot, they found the skeleton, which revealed a head injury consistent with the boy's birthmark. They also found the axe. Faced with this evidence, the murderer admitted to committing the crime.[196]

## Another Fascinating Case

Another spectacular case reported by Ian Stevenson is that of Sujith Jayaratne. This little boy was from Colombo, the capital of Sri Lanka, and he started to display a distinct fear of trucks when he was merely eight months old. As soon as he was able to speak, he said he had lived in Gorakana, a village 10 km from Colombo, and had died after being hit by a truck. Sujith made many statements about this previous life. His great-uncle, who was a monk in a temple not far from this village, heard some of Sujith's affirmations and shared them with another monk who was captivated by this story. He met Sujith, who was then a little over two and a half years old, to discuss his memories. During his conversations with Sujith, the monk took notes, intending to verify the child's assertions later on. According to these notes, Sujith told the monk that he was from Gorakana, and that he had lived in the Gorakawatte neighborhood. His father's name was Jamis and he experienced problems with his right eye. Sujith also claimed that he had attended a dilapidated school and had had a teacher named Francis. The child also claimed to have given money to a woman named Kusuma, who sometimes made him rice noodles. Furthermore, he stated that his house had been whitewashed, that the toilets were located next to a fence, and that he bathed with fresh water. Sujith also said that he gave money to the temple in the forest, and added that there were two monks there, one of whom was called Amitha.

Sujith gave his mother and grandmother more details about this previous life. He told them that his name was Sammy and that he was sometimes called "Gorakana Sammy." His wife's name was Maggie and their daughter Nandanie. As for Kusuma, she was his little sister's daughter. She had long thick hair, and also lived in Gorakana. Sujith added that he had worked for the railways, and had climbed Adam's Peak, a high mountain in the center of Sri Lanka. He had also transported *arrack*, an alcoholic drink whose trade was prohibited. One day his boat capsized, and he lost all his cargo. The day he died, he had had an argument with Maggie, who had stepped out of the house. He had then decided to head to a shop. As he was crossing a street, he was hit by a truck and died.

The monk who had taken notes about his conversations with Sujith went to Gorakana to try to find a family in which there was a deceased member whose life matched his claims. After a while, the monk discovered that a 50-year-old man named Sammy Fernando – and who was sometimes called "Gorakana Sammy" – had died after being hit by a truck. This had taken place six months before Sujith's birth. All of the child's claims turned out to be true, except that he did not die immediately after being hit by the truck. Indeed, Sammy Fernando had died a few hours after being taken to the hospital. Later, Sujith successfully identified several individuals, part of Sammy Fernando's family circle, as well as changes made to his home.

During his investigation, conducted one year after the identification of the previous person, Ian Stevenson spoke with 35 people, including Sujith, who was still talking about this past life at three and a half. Stevenson identified two residents, from Sujith's neighborhood, who had been linked to Sammy Fernando. One of them had been his drinking companion, and the other was his little sister.

In addition to his truck phobia, Sujith had other behaviors that seemed to correspond to Sammy Fernando's life. For instance, he pretended to drink arrack and role-played as a drunk. He also tried to get this liquor from neighbors and smoke cigarettes. In Sujith's family, no one drank arrack nor smoked cigarettes. On the other hand, Sammy Fernando had obviously overindulged in this liquor and in cigarettes. Sujith also demanded spicy foods that Sammy loved and consumed regularly – foods which Sujith's family rarely ate and who would not have thought of giving to a child. Moreover, Sujith had a tendency to swear and to be physically aggressive, which was also the case of Sammy Fernando when he was drunk.[197]

## Reincarnation: Criticisms and Assumptions

In his book *Life Before Life*, researcher Jim Tucker analyzes various objections and possible explanations about reincarnation. One of the objections is the following materialist argument: we should not even consider the possibility of reincarnation because we do not know anything about the mechanisms that can explain it. Therefore, we do not know how consciousness could survive without a physical body, or how it could influence a developing fetus. We need only look at the history of medicine to see the weakness of this argument: indeed, doctors have begun to use some treatments well before their underlying mechanisms were discovered.

Another common objection is that population growth makes reincarnation impossible. In other words, given that today's population is much larger than in the past, all the individuals currently living on planet Earth could not have reincarnated many times in the past. However, some people who are alive now may have had previous lives, while it would be the first for others. New beings could also be created. Another possibility is that some may be reborn because they have things to learn or settle, while others may choose not to be reborn.

In dementia such as Alzheimer's disease, the cerebral degeneration that accompanies this syndrome is associated with memory loss and personality changes. For materialist scientists, this indicates that an intact brain is needed for awareness and that, therefore, memories and personality traits cannot survive death. As Jim Tucker aptly points out, this does not mean that the brain is the source of consciousness, memories, and personality. In fact, instead of producing them, the brain could only allow their manifestation.

When children's claims are not very precise, imagination and coincidence could be the explanation. However, research shows that children providing details about past lives are not more likely to fabricate lies than other kids.[198] Moreover, when the investigated children mention very specific details, or exhibit behaviors (for example, phobias) or birthmarks, the reincarnation hypothesis becomes the most plausible.

There still isn't any theory to lead to an understanding of all the aspects related to reincarnation. However, the most restrained hypothesis to explain researchers' discoveries in this area is that mind and consciousness are not confined to the brain; they would exist before our birth and would continue after our death. In addition, mind and consciousness could attach successively to various physical vehicles (bodies), and lead to memories, emotions, and even traumas in some cases. Hence, the child would identify himself with the previous person because they had been one in the past; memories, behaviors, tastes and preferences, attachments and dependencies, and even physical injuries would be transmitted by the consciousness that survived death to the being having started a new life.

Some prefer the super-psi hypothesis to explain the discoveries of researchers in the field of reincarnation. According to this hypothesis, the children studied by these researchers can gain knowledge of all that is possible to know through extrasensory perceptions. However, it does not explain birthmarks, nor why these children have such a strong conviction of having been someone else. Furthermore, the super-psi hypothesis is hardly reconcilable with the fact that, in general, these children do not manifest any particular psi abilities.[199]

## Hypnosis and Past Lives

Hypnotherapists, who are searching for the source of their patients' problems, investigate their past with the help of age regression. This approach aims at identifying old events that are the source of the problems (for example, particular phobias, dependencies, psychosomatic symptoms) experienced by patients. Sometimes the patients' subconscious produces images and evokes emotional reactions that seem to refer to events that do not belong to the current life of these patients. Therapists using hypnosis obviously cannot determine with certainty whether the contents that emerge from the consciousness of patients constitute a symbolic expression of the *psyche* (like a dream or daydream) or represent episodes related to one or previous lives. Indeed, it is not possible to distinguish between what comes from real memory or the abilities of the mind to reconstruct our autobiographical memory – the memory of the events lived with their context – and the mental capacity to generate false memories and purely imaginary contents. That is why many psychotherapists are wary of hypnosis. Be that as it may, regression hypnosis often leads to the resolution of patients' problems quickly and effectively. Moreover, in some cases, the exploration of past lives makes it possible to find specific details and go back to characters with a historical reality. The hypnotherapist Tatjana Radovanovic Küchler reports such a case:

*Annabelle was simply interested in experiencing regression in a previous life. In a state of relaxation, she saw herself as a married woman waiting for her husband's return in the 1940s. But while waiting, an officer came to announce the much-dreaded death of her husband, aboard a submarine or boat accident apparently.*

*After the session, Annabelle researched the family name she had in the previous life and the date of the accident, which led to a person who actually lived in Canada in the 1940s and whose description matched what she had seen in her regression. This disturbing experience served to convince her that she was truly an eternal soul. She also became certain that her brother, to whom she is very close, was none other than her husband who had died accidentally in the previous life.*[200]

This case, and others like it, are not conclusive evidence of reincarnation. However, as I will show in the next section, spiritual hypnosis is another source of evidence supporting the Reincarnationist Hypothesis.

On the other hand, with hypnosis, Dr. David Cheek, an obstetrician, helped individuals retrieve intrauterine memories that were subsequently confirmed. One of these people, a little girl, remembered that her mother was wearing a dark green plaid dress and had said: "It has to be a girl!" The girl's mother confirmed this information. Cheek believes that the fetal experience depends in part on extrasensory perceptions of the mother and her environment throughout pregnancy. The work of this doctor goes against the idea that babies, at the time of birth or even before, can store information in their memory.[201]

Similarly, Wendy Anne McCarty, a pioneer in prenatal and perinatal psychology, suggests that the baby's abilities to develop self-awareness and a sense of identity involve not only brain development, but also a transcendental self that is not produced by the physical world. This transcendental self is non-local and would ensure that the reincarnating entity already possesses a sense of identity. This self would also perceive the baby's body as a vehicle, which he would use during the incarnation. The human ego, for its part, would be anchored in neurobiological processes. The combination of the two selves would constitute an integrated ego.[202]

## Life Between Lives

Michael Newton was one of the pioneers of spiritual hypnosis.[203]

In the late 1940s, this clinical psychologist began to use hypnosis to help his patients solve the problems that led them to him. His patients often ended up in what appeared to be past lives. Having received a scientific education and being an atheist, Michael Newton did not believe in previous lives, which he conceived as imaginary constructions of the subconscious, and he was not willing to accompany his patients in these said past lives. However, he discovered that he got positive clinical results when his patients understood the links between their problems and events that occurred during the supposed past lives.

One day, during a hypnosis session, one of the patients told Newton that she thought she was in a place and a moment between two of her earthly incarnations. Soon, other patients reported similar experiences. Some patients related what happened at the time of their death, others talked about what they experienced when they found their "soul family," or preparations for a new incarnation on Earth. Michael Newton was intrigued, and he decided to systematically regress all the patients coming to consult him to this life between lives. To do this, he induced his patients into a state of appropriate supra-consciousness, which helped to unlock the memories linked to the spiritual dimension, and used a questionnaire.[204]

Michael Newton thus became a *"spiritual regression therapist,"* a hypnotherapist specialized in life after death. With the protocol he developed, he collected data from thousands of patients for more than 40 years. This would allow him to draw a picture of what happens on the "other side" between two incarnations. The data he collected shows that patient testimonials are consistent regardless of whether they are atheist, religious, or agnostic. For Newton, these testimonies suggested that it was more useful for his patients to rediscover the state of life between lives than their previous lives because in this space-time, they could reconnect with their deepest essence, their soul, as well as their wisdom and all the information that related to them.

Through the data he collected over several decades, Michael Newton was finally convinced of having traced the different stages of the journey of "souls" in the afterlife. This journey begins with physical death and ends with reincarnation when returning to Earth. I present here a summary of what Newton discovered about the various stages of this journey.[205]

At the moment of death, the soul of the person who has just died realizes that it is floating around its physical body. It feels relaxed and curious, rather than fearful – free, happy, and euphoric. It is sometimes frustrated when it realizes that it cannot communicate with the living, who seem to ignore it. Some souls choose to stay close to their bodies, after their physical death, until their funeral as they wish to comfort the loved ones before continuing their journey. They are not upset because they know they will see their loved ones in the hereafter, as well as in future lives.

The soul of the deceased person sees a light of dazzling whiteness far away, behind a dark area which is pulling the soul towards it. This dark area, which constitutes the door of the beyond, looks like a tunnel. The perception of musical vibrations can accompany the crossing of this tunnel. This music heightens the feelings of well-being and calmness experienced by the soul. At the end of the tunnel, the soul of the deceased finds loved ones (soulmates, relatives, friends, spiritual guides) full of love who welcome it to a spectacular place. It then realizes that it has left behind a heavy and cumbersome physical body, and that it must now accept its new state.

At that moment, the soul of the person who has died feels like it is coming back home. The reunion with its spiritual guides and its soulmates who had already left Earth, and with whom it had the most affinities, make it very happy. The guides and soulmates, who are waiting for us near the doorway into the afterlife, are there to reassure the newly arrived soul and help them adapt to its new state. They can manifest themselves as a mass of energy. They can also adopt human traits and choose to appear in an aspect identical to the one they had in past lives with the soul back in the spiritual world. As the soul is androgynous, they can appear in a feminine or masculine form, according to their preference.

The reunion is followed by a transition stage during which the soul receives a "healing bath" to recover from the diseases and traumas of the earthly life it has just left, as well as the trauma of leaving relatives and this life. During the next stage, that of orientation, the soul is invited by one of its guides to assess the life it has just completed. All memories are then accessible, and the soul analyzes everything it has done according to a program it had chosen before incarnating.

As a result, the soul will join its group of souls, which is in fact its real family. The groups of souls (from 3 to 25) are composed of entities of roughly the same spiritual level, who do not leave each other. There are also meetings of primary groups, usually with more than 1,000 souls. All these souls participate in moments of exchange or sharing.

Spiritual guides are actively involved in the realization of our destiny. They remain close to us during our many incarnations, to help us follow our life plan. There are guides of different levels. The quality of the guide that is assigned to the soul is determined by the level of the soul. There is one guide per group. Some very advanced guides work with groups of souls found on both the Earth and the spiritual world. These advanced guides are assisted by beginner guides.

After assessing its past experiences and getting enough resources in the afterlife – since physical lives can sometimes be quite grueling – it is now time for the soul to choose a new life. This choice, which is made in collaboration with the spiritual guides and the group of souls, is made according to what the soul can learn from the point of view of its evolution. Three questions arise for the soul to prepare for this new life: 1) Am

I really ready to face a new physical life? 2) What lessons do I need to learn to increase my knowledge and evolve? 3) What environment and identity would allow me, in my next life, to work as hard as possible to achieve my goals? The soul's project is submitted to a council group consisting of evolved entities in the place of selection of lives. The members of this council, whose role is not to judge but rather to help, provide information to souls who want to be reincarnated to help them work on negative behaviors in the next incarnation. The place selected for the next physical life is like a movie theatre: souls can see scenes of their potential future lives unfold there. But it is not possible for them to see and know everything. This makes souls come back to Earth, or incarnate on another planet, with a kind of global map, or life plan, based on what they have chosen.

The choice of body is also of major importance. Michael Newton's patients reported that this choice is influenced in part by karma, the principle of causality according to which all our thoughts, words and actions are causes, and all our experiences their effects. Newton's patients also revealed that selecting difficult lives and far from perfect physical bodies (for example, with disabilities) allow faster progress on the spiritual level. Another important aspect of choosing a new incarnation is the selection of the psychological traits of the character in which the soul chooses to incarnate. These traits must allow the soul to reach its life goals.

The choice of parents and family is obviously another essential part of the new life. Some people find it extremely difficult to accept this idea because they are convinced that they are innocent victims who have had the misfortune of being the offspring of bad parents. However, as souls possess free will, they are never victims without the qualities and tools to overcome hardships and fully realize their life programs.

The soul that wishes to incarnate must also "synchronize" with other loved ones – for example, siblings, friends and spouses

– who will participate in its life scenario. This synchronization is mutual insofar as this soul plays a significant role in the lives of all those close to it by allowing them to realize their own life plans. In this context, it is imperative for the soul that incarnates to be able to recognize the members of its group of souls with whom it must interact in its future life. It thus goes to the place of recognition where signs are encoded in its memory. These signs, which are associated with the five senses (for example, a particular way of laughing at a person or a piece of jewelry worn by another person), will create memories that will allow the recognition of these members to key moments in the life of the incarnated soul. As for meetings with a soulmate, who will eventually become a male or female companion, an intense moment full of "magic" allows recognition.

When the soul is finally ready to reincarnate on Earth (or elsewhere), it greets its friends and is escorted by its spiritual guides into a boarding lounge for the return journey. It then takes a tunnel and comes to merge with the body of a child in pregnancy, in the mother's womb. Some souls come to inhabit the body of the embryo very soon after conception but usually they reach the host to which they are assigned during the fourth or fifth month of pregnancy. In the fetal state, they become accustomed to their host's brain while remaining conscious of being immortal souls. After birth, amnesia sets in and souls combine their immortal characters with the temporary human spirit of the host. It is this combination of traits that creates a new personality.[206]

## Chapter 8

# Journeys to the Source

*The total number of minds in the universe is one... In truth, there is only one mind.*
Erwin Schrödinger

## Other Transformative Experiences

Like Near-Death Experiences (NDEs) and the phenomena discussed in previous chapters, spiritual experiences (SEs) represent another category of manifestations that contradict the materialistic scientific ideology. SEs are usually experienced in an expanded state of consciousness. During a SE, the experiencer accesses a "larger" reality, which transcends the material world, and can also come into contact with nonphysical entities.

SEs are more common than we think. For example, a survey conducted in 2004 by the National Opinion Research Center (NORC) found that 50% of Americans had an SE that had greatly impacted their lives.[207] These experiences often take place during prayer or meditation, or while contemplating nature. SE can also accompany physical or mental health problems, including depression[208] and those who experience them can change dramatically. Thus, following intense SEs, perception of the world widens, and life acquires a new meaning; beliefs, values, attitudes, behaviors, and the sense of identity (the sense of self) are transformed. Kindness, compassion, unconditional love, and helping others become important. Experiencers realize that they are spiritual beings continuing to exist beyond the physical plane. They have a deeper sense of interconnectedness with Nature – the existence of a transcendent principle (which many call "God") becomes a reality, which helps experiencers to face problems with greater serenity. Experiencers usually acquire a whole new perspective on their problems, and they no longer feel victims of day-to-day hardships. At times, SEs lead to physical or emotional healing.[209]

Here is the story of a remarkable SE that a man living in the United States shared with me:

*During an evening in 1977, I was in my car, ready to go home when suddenly, faster than a thought can become a thought, I found myself in a world of Angels without any trace of material reality [...]. I found myself facing a Glorious Guardian Angel, enveloped in a white robe of Light. I was also wearing a similar robe. We looked into each other's eyes and stared at eternity. I realized, at that moment, that this world was eternal, as well as this angel, myself and all conscious beings. The Guardian Angel disappeared, and suddenly, all my incarnations flashed up on a cosmic screen. I then understood the material life illusion that imprisons us. I cannot say that I emerged from this experience a changed man because, in truth, I did not come back at all except for some fragments of an old life. I severed all ties with everyone I knew and with the life I led. I no longer lived as I did before this experience, when I was only concerned with myself. After my experience, I cared about everyone because I knew from now on that my essence was the same as everyone else [...]. I feel that my brain was transformed during this experience because since then, my mind is always in the present – neither in the past nor in the future.*
*– Vincent R.*

There are different types of SEs. One of these is the mystical experience (ME). MEs are at the root of most major religions. These experiences, which allow the experiencer to become aware of cosmic realities that cannot be grasped during regular states of consciousness, have thus greatly influenced the history of humanity. MEs have the following characteristics: loss of sense of self; outside time and space; impression of having reached the ultimate foundation of reality; feelings of peace, happiness and joy. ME, which cannot be expressed in conventional language, is also a source of non-rational and intuitive knowledge.[210]

During certain MEs, there is a temporary union, even an identity, with the ultimate Reality, which is identified in different forms (the One, the All, the Absolute, the Ground of Being, the Source, the Universal self, God, Brahman, Tao, Allah, etc.). Individuation then disappears and is replaced by the perception of being one with all things in the universe. The experiencer enters at that moment into a state which is called in certain traditions (for example, the Advaita Vedānta) "non-duality." This state is accompanied by the intuition that the Ground of Being is at the origin of all existence and that it expresses itself through an infinity of distinct forms. Psychiatrist Richard Maurice Bucke coined the term "cosmic consciousness" to refer to this subtype of ME.[211]

Through virtually all great spiritual traditions and eras, historical figures have reached the state of cosmic consciousness, either temporarily or permanently. However, what is remarkable is that the experience of cosmic consciousness can also happen suddenly to individuals who until then had shown no interest in spiritual things.[212] The following account illustrates this possibility. It was sent to me by a woman named Sheila:

*As I was a firm atheist, I paid little to no attention to everything that seemed even remotely related to religion. But my skepticism to religion could not prevent my spiritual awakening [...]. On February 14, 1997, my perception of the world suddenly changed dramatically. Abruptly and without warning, an element of my consciousness, my soul, escaped from the confines of my body and rose above the bed on which my body was resting. Shortly after, the material plane around me seemed to disintegrate into a sea of particles and then in a foggy ether. Then I crossed a sort of threshold separating the material world from the metaphysical world to float in an oceanic universe [...]. I noticed then that my breathing was in tune with a breathing sound echoing through the universe. This cosmic breathing seemed to come from the ether around me and seeped into my body found in the room. It is at that moment that I became extremely attentive to the presence of a living and conscious entity that permeated the entire cosmos. Following this discovery, my sense of self faded to contain this cosmic Consciousness [...]. During this unitive experience with God (the Creator), the veil that obscured my thought rose [...] until my soul hastily left the metaphysical plane in order to reintegrate my body [...]. Discovering the spiritual dimensions of my human condition for the first time was like putting on a new pair of glasses. My perspective changed abruptly [...]. I morphed into a new creature: a mystical one. I now saw the world through different eyes. My thoughts changed radically [...]. My sense of identity was no longer bound to the content of my mental activity. I did not define myself in terms of my culture, my gender, my beliefs and the color of my skin anymore [...]. I now perceived myself as a spark of Cosmic Consciousness incarnated in a member of the human species.*

It is conceivable that some individuals in search of the Absolute – and having spent a large part of their lives praying, meditating, and contemplating – could reach the state of cosmic consciousness, during humanity's great adventure. But that people who have no interest in the spiritual can also reach such a state is much more difficult to understand and explain. In these cases, which seem rather rare, it is possible that the experience of cosmic consciousness is given by the spiritual essence of the experiencer – the deepest part of his being – in order to transform his life and guide him towards the realization of his highest potential as a human. It would be part of the evolutionary process of consciousness.

## The Rise of Kundalini

Another type of ME is associated with kundalini, a yoga Sanskrit term for a powerful energy curled up at the base of the spine at the level of the sacrum. Represented as a self-winding serpent, kundalini is referred to as "cosmic energy," "creative energy," or "divine energy," according to the various traditions that use this concept. Only a small part of this energy would flow into the body of most individuals. For some yoga schools of thought, the purpose of different spiritual practices (e.g., meditation, breathing, postures) would be to awaken that energy. Once awake, the kundalini would be felt as ascending along the spine, crossing and activating, one after another, the energy centers (or "chakras") of the human body. When this energy reaches the top of the skull at the level of the fontanel, it would follow a state of expansion of consciousness leading to the impression of "one" with the All.[213]

The concept of kundalini is not only found in Hinduism, since it appears under other names in Buddhism, Taoism, and the Kabbalistic tradition. Some Christian theologians have also drawn a parallel between kundalini and the Holy Spirit.

The "rise" of kundalini is frequently accompanied by perceptual manifestations (for example: flashes of light, buzzing sounds, sensations of heat, tingling, and vibrations). Feelings of fulfillment and bliss often accompany the awakening of this energy, which can be felt as a process of inner purification. However, the awakening of kundalini can reactivate deep fears and former traumas. Moreover, if the intensity of the awakening of this energy is too great, the symptoms can become unpleasant and the experience of the kundalini can lead to psychological disturbances.[214]

Here is the summary of a kundalini experience that I myself

experienced a few years ago:

*I was showering at around 11:45 a.m. During the shower, I felt a boiling sensation in my solar plexus accompanied by nausea. I got out of the shower. I had the strange feeling of losing my identity (being a neuroscientist, I told myself that these symptoms could be related to a tumor). Curiously enough, I was not worried. I was in an altered state of consciousness (as if I were drugged). Mechanically, I got dressed and went to bed at around noon. Powerful waves of energy then surged through my body, from the solar plexus to the heart and to the head. My heart was beating rapidly; I was panting. Again, I wondered if what was happening to me was pathological or dangerous. I heard a voice telling me: Don't be scared. There is no danger. Let yourself go.*

*I then managed to abandon myself completely. The waves of energy became ecstatic and enveloped me with absolute and unconditional love. Eyes closed, I perceived a golden white light shining and loving. I rested in this Light. I merged with the Light and my identity disappeared [...]. I realized that I could jump into the Absolute and surrender to the Source of all things, that everything was perfect, that you must never worry about anything. During this fusion with the Light, I perceived that all that I needed in order to fulfill my mission of life would be provided to me, that I would be constantly guided. I would receive the necessary information for achieving my future projects [...]. At the heart of the experience, I still perceived my body, but I felt at the same time very far from the Earth. I could have died physically and that would not have bothered me at all. Gradually, the force of energy waves decreased, and the rhythm of my breathing stabilized. Gently, I felt myself reintegrating my body. I opened my eyes and then looked at my watch: it was 12:30.*

## Shamanic Experiences and Entheogens

Shamanism refers to the millennial practices and rituals that are still present in several cultures (for instance, in Siberia, Mongolia, Nepal, Africa, Australia, and among Native North and Latin Americans). In shamanic cultures, initiates or "shamans" resort to various techniques that alter their state of consciousness. These techniques – which include fasting, sweat lodge, drumming, rhythmic hyperventilation, meditative approach, and taking entheogens (psychotropic substances inducing an expanded state of consciousness used for spiritual or shamanic purposes) – mix music, dances, and songs. The trance states thus induced are supposed to allow shamans to come into contact with the dead, the deities, or the spirits of nature (animals and plants). This form of contact would enable shamans to experience mystical states of consciousness, or access information that gives them the ability to heal diseases or predict the future.[215]

In Western countries, more and more people on a spiritual quest feel attracted to shamanic practices and go to distant lands to meet shamans and participate in initiation rituals. Surprisingly, individuals who do not know shamanism or are not interested in it sometimes report "shamanically-sounding" experiences. In such experiences, these individuals may feel as if they are merging with the body of an animal. In shamanism, this phenomenon is called "shamanic fusion."[216] The following account is a good example of such an experience:

*I experienced my first fusion more than 10 years ago, and despite that, it remains present in my memories as if it had just happened. I also remember it as one of the most beautiful experiences I have ever had. The trigger was, however, purely accidental. At the time, I had little knowledge of fusions, so experiencing one was not my goal. I used to frequent an esoteric bookshop in the Liège region whose distinctive feature was to play a wide variety of New Age or relaxing music. Obviously, the music was varied and one day Rudy, the vendor, had decided to play a CD by Phil Thornton titled*

Initiation. *Since there were only two of us in the store, the music was clearly audible, and Rudy had increased the volume.*

*I found myself in front of the counter, just listening until a white flash propelled me into this desert of red-orange sand scattered with bushes. For a moment I thought: "How did I get here?", the next moment I didn't care. I felt that warm wind on my face; at that moment my place was there. I turned slowly and saw this mountain lost in the middle of nowhere. It looked like a giant had snatched the top and I knew that I had to walk towards it. I finally reached the foot of this immense red monolith. I was as happy as if I had returned home after a long trip or a long absence. I stared at the rock for several minutes; I looked at these veins and the cracks running through the stone, I found it wonderful... Then I heard the voice, I turned in its direction. The landscape changed, the plains became dunes and on one of them stood a cougar. I don't doubt for a moment that in real life, I would have fled as fast as I could. But the cougar, this cougar – words were not enough to describe it. I walked towards it and joined it on the dune. I looked into the eyes of this animal that stood there in front of me and I doubt that I will ever be able to find the words to describe what I saw there. The voice spoke to me once more and intuitively I knew it came from the cougar. These words were:*

*"Here you are, finally, little brother of the Earth."*

*I remember thinking: "It's impossible, animals don't speak," and the voice answered me:*

*"Indeed, I do not speak to you with words. Why speak with words when it is more beautiful to speak with the mind?"*

*"With the mind?" I replied. His answer was:*

*"Soon you'll understand."*

*I didn't know what to think, I told myself that I had to be dreaming. I placed my hand on him and he appeared more real than ever. I asked him:*

*"Why am I here?" And I already knew the answer he would give me.*

*"To learn, little brother. To live, little brother."*

*Questions were jostling around in my head.*

*"Why do you call me little brother?" He answered me:*

*"Under this skin there's flesh and under this flesh there's blood just as under your skin. In my chest beats a heart, just as in your chest. If you pierce this flesh, like you I feel pain. The blood that flows will not be less red than your blood. Break this heart and like you, little brother, I will suffer. Come, let's walk!"*

*We walked for a while and went back to the plains because it was hard for me to follow the cougar in the sand. We walked in silence, in the beauty of silence. Back on the plain, the cougar told me:*

*"Come on, little brother, let's run! Let the wind caress you, listen to it sing..."*

*We started running and I was quickly left behind. The cougar stopped to wait for me and once I joined him, he asked me to forgive him for running faster than me.*

*"You're not capable of running like a cougar, you're still young in your learning process, I should have taken this into account. You cannot run like a cougar, so you have to become the cougar..." he told me.*

*I felt myself becoming light, misty.*

*A few moments later, I was inside the cougar. I had merged with him, I was in him and I saw through these eyes like a spectator. I could still hear his thoughts. He explained to me that he had fused our two beings, and that in due course, he would teach me how to achieve this too. The cougar resumed his run and we ran a long time on the plain, although I was only a spectator in this body which did not belong to me, I could feel muscles (which I did not have) working, fill with blood, contract and soften as the heat spread to these muscles. I found this feeling to be wonderful.*

*We heard a scream; the cougar stopped and looked up. An eagle hovered high in the sky. The cougar explained to me that it was another of my brothers, a bird of power as he himself was an ally, an animal of power. I thought it must be as wonderful to hover in the immensity of the sky as to run on the plains. As if he was still openly reading my mind, the cougar then told me, amused:*

*"Find out for yourself!"*

*I felt that I was leaving the cougar's body and my spirit was rising towards the eagle and I ended up fusing with the eagle as I had done with the cougar. My perceptions changed again, and I could see my feline friend running in the plains. I felt the wind caress the feathers; I even felt the currents ascending and descending… I was flying! The eagle asked me if I liked these sensations. I was so drunk by the speed and the multitude of new feelings that I did not know what to answer.*

*Sadly enough, we ended up landing not far from the cougar.*

*"Unfortunately, we each have to go our way, little brother."*

*I was terribly sad to hear that, and he must have felt this sadness because he added:*

*"Do not be sad, it's not an ending, or a goodbye. It's a beginning; the road is long, we have taken only one step and there are a thousand left." I asked him:*

*"When can I see you again?"*

*He replied:*

*"I'm always by your side, whether you can see me or not. See you soon."*

*Before I had time to say anything, I found myself in front of the counter again facing my friend Rudy. He knew that something had happened and told me that I had seemed completely absent for quite some time. He said that he had felt he was in front of an empty vessel, facing a soulless body. He had even been worried that I would faint. I was experiencing some kind of emotion and I started crying with joy, stammering in the eagerness of what I had to tell him… But tell him what? A dream? A vision? What exactly had I experienced?*[217]

Spiritual and shamanic experiences can also result from taking entheogens (this term means "generating the divine on the inside"). I can personally attest to the fact that at the subjective level, the entheogens – also called psychedelics – lead to spiritual experiences similar to mystical states associated with spiritual practices or occurring apparently spontaneously. For Stanislav Grof, a psychiatrist who has worked with these substances and is one of the fathers of transpersonal psychology, entheogens (e.g., mescaline, psilocybin, ayahuasca, and LSD) appear to act as catalysts/amplifiers allowing deepest levels of the *psyche* to manifest.[218]

Entheogens may induce experiences in which the little self (which some philosophical and spiritual schools call the "ego") of the experiencer is transcended and this experiencer merges with an infinitely wider Consciousness which is also full of Wisdom and Love. These inexpressible experiences, which appear more "real than reality," are frequently accompanied by feelings of intense peace and happiness. In the long run, entheogenic experiences can reduce egocentric motivations (e.g., desire for power and fame) and foster a personal transformation that includes social and ecological awareness.[219]

One of the most popular entheogens is ayahuasca. This drink concocted from lianas – which contains DMT, a powerful psychotropic drug – has been used for a very long time by the shamans of the Indian tribes of Amazonia. Generally, ayahuasca consumption is portrayed by its users as an intense and positive experience that allows for a new understanding of life, the world, and the self. The use of this entheogen also improves interpersonal relationships and promotes a change in values as well as a new relationship with nature. Taking ayahuasca also promotes spiritual development.[220]

The following is the story of a Westerner, Marie-Claude, who went to the Amazon rainforest to experience this entheogen under the supervision of a shaman named Ricardo:

*All fears fled when I met Ricardo, a man full of wisdom and joy; trust was established. His radiant face seemed to me full of certainty, so what good was it to fight. I knew that nothing could happen to me if not a wonderful journey still clinging to me, with its roots deeply set inside me. The sacred ceremony could begin, I trusted him completely and I knew that my only reason was to live this precious moment in the abandonment of myself.*

*From another time, with the seal of his gaze on me, I was already beginning to center myself. A moment of well-being in the sharing of the circle. The ritual could begin, I knew that another reality would be offered to my soul [...]. It was the plant's turn (ayahuasca). Its bitter taste seeped into me, and there, I began to speak to it, asking it to be sweet with me and to lead me to my reality and that, thus, I would follow it. In fact, I wanted it to show me the way and that's what it did, and this is how. In harmony with the sweet melody of Ricardo's voice [...], it was not long before I departed [...] – it was directing me, and I felt it immediately throughout my body. It passed with extreme rapidity along my spine, making me feel every vibration, every bone; life vibrated within me. Everything resonated in me. I felt my organs; each passage released a pent-up tension. I had the feeling of being delivered from the weight of constraints. I felt light. My body did not belong to me anymore, or at least, I did not control it any more. The plant had literally tied me up, I was its prisoner, I could not move. I told it OK, that's what you want then I'll let you be the master of my soul, I'll be your student, teach me. Then, suddenly, the speed increased, a bright field offered itself to me [...] The lights [...] filled my head, bursting inside me [...].*

*I will not dwell on my vomiting bouts (so much to get out of my body). The doubts, the fears clogged in my unconscious – this precious well, it had to spit out its venom. No time to dwell on my flaws – because I yearned to learn, know and explore. I had hardly asked that, quickly and efficiently, my wish was fulfilled, and I received the revelation. Stars swept, with an intense luminosity that I absorbed – it seemed as if they came to die inside me and suddenly reappeared through an angle of my vision, a column of hands joined together in a circle [...]. Then on that foggy background stood a cross, so luminous that I could not take my eyes off it. A cobra was curled up around the cross and long plants with extremely thin stems spread out to ultimately hatch into exquisite silver filaments. These filaments urged me to look ever further as if the infinite belonged to me, as if only I could set the limits of my possibilities.*

*For a moment, I heard Ricardo's voice singing, I know I knew a few lyrics to that song, for I remember saying it at the same time as him, as if I was familiar with his knowledge.*

*In my quest, an elderly woman with dark, wrinkly skin, extraordinarily powerful black eyes, a wide mouth, and gaping teeth began to smile at me and talk to me. She motioned me to follow her; at that moment I felt I was leaving my body. I went on a journey with her over a nebulous opaque space, and suddenly a circle centered in an immense rectangle appeared to me; this circle was turning in arabesques. It represented the center of the universe. I plunged into its center and there I discovered a green hill where luminous triangles (three at the top and two at the bottom in staggered rows) showed me the way, their light traced an arrow like the indication of a road to follow, which I did and there again this woman showed me an enormous and very old lock, and said to me: "Open it, you have the key." That's what I did, the door was heavy, but on the other side the light was blinding. There, many radiant entities of light smiled at me with happiness and joy, everything was calm, relaxing, with the extremely colorful and luxuriant nature.*

*Suddenly, I felt the sensation of returning to my body with a jolt and hearing the circle again. I know that at some point I started talking to some of my neighbors, telepathically, of course. Ricardo, for a long time, officiated in the circle, I know that I could not look at him, with his vibratory field so intensely radiant that he was blinding me. Others were releasing their woes – this dual sensation of being both present and absent, but of sharing a communion, of feeling the warmth of this circle. It was wonderful when Ricardo approached me to care for me. I felt his extreme strength and sweetness on my fontanel, my eyes filled with tears and sobs rocked my body. The feeling of being protected, rocked even if all this was accompanied by vomiting, but again, I knew it was necessary, his hand purified me. After that, I could indulge in my visions without restraint. I left accompanied by a crowd of animals – they were all looking at me with their eyes sparkling. One of them, a wolf, crouched in front of me. Nature offered itself without restraint. I explored it and an infinite chatter of trees filled my left ear. It was very irritating because I did not understand their language, but I can tell you that it was incessant; they were real gossips. On the right, I was talking to the entities. Suddenly, I was projected with a breathtaking speed towards the summits in a dizzying ascent. A huge net, with very large, silvery stitches, stopped me. I could get through and at the same time I could not, as if a border was denying me access. I came down again and I felt as if I was in a futuristic city, judging by the size of its houses. Several times, I saw an apartment with a large terrace further back, at the top of a building that was at the corner of a street. Once, I entered an old house. The lobby was large, like that of a castle and the stairs were made of stone and the walls were greyish green. As I climbed, I looked at a gallery of family portraits, and when I reached the landing, they faded from my view to make way for a small door at the bottom of the gallery. So, I started shouting, "No!" Twice I came back to this scene. I was paralyzed. An important thing too – I needed to inhale, to breathe, to fill my lungs with air from elsewhere, but I don't know how to explain that, because I had the feeling that it was vital. To take in, at all costs, this new air, to fill me with it. To be honest, it took me six days to realize the necessity to feed myself. And yet, I love food, but it felt useless. There was the feeling of already being nourished. Other visions came to fill my soul, but the most vivid ones are those that I express here. My body and my mind have long benefited from the effects of the plant, a healthy well-being.*[221]

Although Marie-Claude's experience reported here was very positive overall, I do not want to give the reader the impression that the use of entheogens is completely safe. In fact, taking these substances can cause reactions of fear or panic (*bad trip*) throughout their active duration. Using entheogens can also cause long-term perceptual disturbances, as well as precipitate or exacerbate symptoms associated with mental disorders.[222]

## Spirituality and Health

Unsurprisingly, materialist neurologists and neuroscientists do not hesitate to "pathologize" spirituality and SEs. Some of them have stated that electrical disturbances of the temporal lobes of the brain – located on either side of the skull, just above the ears – play a significant role in SEs. According to this hypothesis, the temporal lobe would act as the "module" of SEs in the brain, and epileptic seizures located in this structure could induce this type of experience.[223] This hypothesis suffers from some major weaknesses: on the one hand, surveys conducted in several countries reveal that SEs are frequently reported by individuals who do not suffer from an abnormality such as temporal lobe epilepsy (TLE). On the other hand, it is extremely rare for people with TLE to report SEs. If TLE actually causes SEs, all those who suffer from this neurological disorder would have such experiences.

This does not mean, however, that the temporal lobe is not involved in the cerebral mediation of spiritual and mystical experiences. In this regard, some neuroimaging studies conducted in my laboratory suggest that the temporal lobe actually plays a role in the perception of contact with a spiritual dimension. The results of these studies, however, indicate that other brain regions involved in perception, positive emotions, representation of the body in space and self-awareness, are also correlated with SEs. Therefore, there is not a single "point of God" (or module of spirituality) in the temporal lobe.[224]

Neurologists and neuroscientists are not the only ones who have attempted to pathologize SEs. During the second half of the nineteenth century and much of twentieth century, some psychiatrists and psychoanalysts tried to do the same by suggesting that SEs were symptoms associated with mental disorders resulting, for example, from abnormal psychological defenses or an immature personality.[225] However, contrary to what these psychiatrists and psychoanalysts believed, a growing body of scientific evidence shows that SEs and spirituality have several beneficial effects. Indeed, these data indicate that a higher level of spirituality is associated with better mental health, greater well-being, better quality of life, higher self-esteem, more optimism, and higher longevity. In addition, a higher level of spirituality is associated with lower rates of stress and anxiety, depression, alcoholism, substance abuse, and suicide.[226]

## The Inner Connection

My research about the neurobiological correlates of spiritual and mystical experiences has received international media coverage. In this context, journalists have often asked me how the identification of brain regions associated with such experiences could be useful to the average person. After being asked that very legitimate question a few times, I started to reflect on how to allow people who are not experts in contemplation to reach a state of consciousness allowing them to get in touch with their spiritual essence.

During my reflection, it occurred to me that brainwave entrainment could be an interesting way. This concept refers to the fact that the frequency of brainwaves can synchronize, through a resonance effect, with the frequency of an external stimulus, for example a sound.[227] In order to change the frequency of brainwaves, we can use so-called "isochronic" sounds. These sounds are regular beats of a single tone. In other words, an isochronic sound is a tone that fades and reappears quickly at regular intervals. This creates distinct and precisely defined sound pulsations, which are similar to the beat of a drum. These isochronic sounds can be used to accelerate or slow down the frequency of brainwaves.

When we change the frequency of brainwaves, we automatically bring a modification to the state of consciousness and a momentary change of the sense of identity. Based on this principle and some pilot studies carried out in my laboratory, I decided a few years ago to embed isochronic sounds of specific frequencies in some musical frames designed for relaxation and meditation.[228] These preliminary studies have shown that SEs and contact with spiritual planes are associated with a significant slowdown in the frequency of neuroelectric activity relative to normal waking consciousness.[229] These studies also revealed that infrasounds, sounds whose frequency are less than 15-20 Hz, which are not perceptible consciously, can modulate the frequency of brainwaves.

Using this technique of nesting isochronic sounds in musical frames, I created a psychospiritual approach titled *"The Inner Connection: A Conscious Re/Creation."* In the initial phase of developing this approach, I used this technique to help, individually or in small groups (in the form of workshops), individuals interested in personal and spiritual growth to enter into a broader state of consciousness. Isochronic sounds, whose frequencies correspond largely to those of slow waves (theta and delta), were used.[230] The results of this work have shown that at the sense of identity level, this expanded state of consciousness allows a passage of "little self" (also called the ego or the superficial self) to a "Big Self" (also named Higher

Self or Ego). In the state of the Big Self, which corresponds to the spiritual part of the *psyche* (or supraconsciousness), it becomes easier to free oneself from limiting beliefs, negative emotional patterns, and inner wounds. The state of the Big Self also allows us to intuitively receive information that can be related to, for example, the life plan or what appears to be "past lives." This state also allows us to experience spiritual love and to generate new experiences and realities.

Frequently, individuals in this state realize that everything is fine – even what appears to their little self as serious problems (for example, a life-threatening illness such as cancer) – and that we must trust in life, let go, and surrender ourselves to our spiritual essence. There is also an awareness that the little self is transient and malleable.

With regard to emotion, the state of the Big Self is associated with well-being, peace, lightness, and joy. At the energy level, individuals in this state often feel a subtle energy entering the heart region and/or the "crown chakra." They can also perceive waves of energy circulating in the physical body, which can lead to an ecstatic reaction. Individuals in this state sometimes also claim they meet spiritual guides; in other cases, they have the impression of merging with their Big Self.

# Conclusion: The Emergence of Postmaterialist Science

*The architecture of the universe is consistent with the hypothesis that the mind plays an essential role in its functioning.*
Freeman Dyson

## What We Learned in this Book

I have tried to consider all the research in this book that informs us about consciousness and mind-brain relationships. As the reader will have noted, the empirical evidence in this research reveals that the old materialist paradigm is too restrictive: undoubtedly we need an expanded scientific model of reality.

In the first chapter, we have seen that intentions, beliefs, and emotions greatly influence brain activity, as well as that of genes and the physiological systems (endocrine, immune, cardiovascular, etc.) connected to the nervous system. Hence, mental activity plays a significant role in our health and well-being. All this shows that we are not biological machines ("meat puppets") totally defined by our brains, our genes, and our environment.

Our mental abilities are not limited to the confines of our bodies since, as I indicated in Chapter 2, we can capture information without using our ordinary senses, in ways that transcend the usual constraints of space and time. The studies discussed in the following chapter indicate that intention can even remotely affect the wave function in quantum mechanics (QM), the activity of random number generators and other electronic devices, as well as the structure of crystals of water molecules. In Chapter 4, we have seen that the human mind can also deliberately influence biological systems, such as plants, enzymes, and the physiological activity of others.

In addition, I have demonstrated that individuals may have

true perceptions – corroborated by independent witnesses – during a state of clinical death triggered by cardiac arrest. These individuals can also live deep transcendent experiences while they are clinically dead. These data strongly suggest the continuity of consciousness and personality after death, as well as the existence of nonphysical domains of reality. The evidence presented in Chapter 6 concerning communication with the deceased, induced post-mortem communication, scientific studies on mediums, and instrumental transcommunication reinforce the notion of survival of consciousness after death. Scientific research on reincarnation, and Michael Newton's work on Life Between Lives (presented in the chapter that followed) are also along the same lines.

Finally, in Chapter 8, I discussed the expansions of the field of consciousness that are associated with transcendent experiences. As I have shown, these experiences can occur spontaneously. They can also occur following a "rise" in kundalini, in a shamanic context or in conjunction with the taking of entheogens. Whatever the trigger, these experiences support the idea that we are much more than our physical bodies.

Pseudo-skeptics, who try to pose as genuine skeptics to the general public, will claim that the empirical evidence presented in this book can be explained by physical mechanisms, and that they do not question the materialist worldview. What is essential here is that these pseudo-skeptics have absolutely nothing in common with genuine skeptics. Unbiased, they conduct investigations and research with an open and objective mind, because they are motivated by a desire for understanding and knowledge, as well as the search for truth. True skeptics also take into account all the evidence. Using critical thinking and rational analysis, they question the facts and their interpretations. They are cognitively supple and do not jump to hasty conclusions; they are ready to question their own beliefs, as well as to adjust their hypotheses according to new discoveries.

Pseudo-skeptics, as for them, are fundamentalists engaged in a crusade to defend the materialist doctrine at any price. Thus, these "fundamaterialists" gathered to conduct a guerrilla on Wikipedia: they managed to infiltrate this online encyclopedia. Working in teams and using pseudonyms, they make sure that this online encyclopedia presents research on psi phenomena, as well as some aspects of research on alternative and complementary medicine, as pseudoscience. These fundamaterialists also manipulate the biographical pages of the researchers involved in this research.[231]

Pseudo-skeptics present themselves as defenders of critical thought, logic, and reason. With a closed mind, they are not interested in facts and truth. Thus, they do not seek to examine the evidence or conduct experiments. Moreover, they deny all the evidences that are not compatible with the materialist doctrine or they try to convince that, sooner or later, physical explanations will make it possible to demystify all the empirical data. Intellectually dishonest, they do not hesitate to lie and resort to disinformation, as well as to attack scientists whose work demonstrates the erroneous nature of their sacrosanct doctrine. What is most pathetic is that despite these, pseudo-skeptics claim to know what is true and what is false, what is possible and what is not. Ultimately, they seek mental control of the population.

## The Structure of Scientific Revolutions

In 1962, historian and philosopher of science Thomas Kuhn published a book titled *The Structure of Scientific Revolutions*[232] which became very influential in the academic world and some popular circles. In this book, Kuhn proposes that paradigms – the theoretical frameworks of scientific disciplines within which theories are formulated, and the experiments conducted – can and must change because sooner or later they will no longer be able to explain the phenomena that are observed. Kuhn also demonstrates that scientists are generally unable to recognize the existence of phenomena that are not allowed by the paradigm to which they adhere. In this respect, he cites the following example: it was not until Copernicus proposed a new astronomical paradigm that astronomers in the Western world could see that the heavens were not unalterable. While Chinese astronomers, whose cosmic beliefs did not exclude celestial changes, had noted the appearance of new stars in the heavens long ago.

According to Thomas Kuhn, when anomalies accumulate – experimental observations or other empirical evidence that violate the widely accepted theoretical framework – and the persistent efforts of researchers to elucidate these anomalies fail, the scientific community begins to lose confidence in this paradigm and a period of crisis ensues: a new paradigm, which competes with the old dominant paradigm, can now be considered. This new paradigm, which is not at all an extension of the old paradigm, represents a totally different view of the world.

Typically, daring and visionary scientists champion the new paradigm, and attack the old theoretical framework. Unsurprisingly, as conservative scientists believe that anomalies will soon be resolved by the old paradigm, they are struggling to save the old theoretical framework. However, if the new paradigm is sufficiently promising, i.e., if it is better able to explain the anomalies in question, then a significant number of scientists are moving away from the old paradigm, and a change of paradigm (or a scientific revolution) occurs. Following this paradigm shift, researchers are returning to a period of normal science, and they are trying to solve problems by staying within the new theoretical framework.

The Copernican Revolution – a radical reversal of the representation of the world – is a very good example of a major paradigm shift: Ptolemy's current geocentric model was followed by a heliocentric model championed by Nicolaus Copernicus. This model was then perfected by Kepler, Galileo, and Newton. Remarkably, Aristarchus of Samos, a Greek mathematician and astronomer of the third century BCE, hypothesized that the Earth turns on itself and moves in a circle with the Sun as center. But since the vast majority of astronomers adhered to geocentrism, it was not until eighteen centuries before Copernicus proposed his revolutionary model.

Another example of a major paradigm shift in science is the development of QM between 1900 and 1930. This new physics emerged as mathematical explanations for certain anomalies at the atomic level that were not compatible with the prevailing theories of classical physics.

Current scientists doing research related to consciousness and mind-brain relationships are in a similar situation as that of physicists at the very beginning of the twentieth century. They are faced with an ever-increasing number of anomalies that cannot be elucidated within the framework of materialist theories.

## A Personal View of the Postmaterialist Paradigm

The empirical evidence discussed throughout this book tells us that now is the time to get rid of the materialistic shackles and expand our understanding of reality.

Even though we do not have all the answers to our questions yet, regarding consciousness and mind-brain relationships, we can still present some key elements of the postmaterialist paradigm.

One of these key elements is the following: the mind is irreducible to physical processes, and its ontological status is just as important as that of matter, energy, and space-time. In other words, mind does not come from matter, and it cannot be reduced to something more fundamental. In this regard, the philosopher David Chalmers[233] and the physicist and cosmologist Andrei Linde[234] postulated that consciousness, one of the main functions of the mind, is a fundamental component of the universe.

This element of the postmaterialist paradigm is based on the undeniable fact that consciousness is a prerequisite for reality. Indeed, the physical and mental aspects of reality are subjectively experienced through consciousness. Without it, it would be impossible for us to appreciate the musical genius of Mozart or the exquisite taste of Dom Perignon champagne; and there would be no sense of identity or apprehension of the world. On the other hand, it is becoming increasingly clear that the concept of mind is absolutely necessary in order to understand the universe more accurately, be it at the micro physical level (for example, the effect of the observer in QM) or at a more macroscopic level (for example, mental influence on brain activity and behavior).

If the mind is primordial, then it is conceivable that mental processes and events, including subjective interiority, exist at the various levels of organization in the universe and that they do not absolutely require a nervous system to operate. In this respect, theoretical physicist and mathematician Freeman Dyson proposed that since atoms behave in the laboratory as active agents rather than as inanimate matter, they must possess the ability to make choices.[235]

At the molecular level, there is evidence to suggest that molecules composed of only a few simple proteins are capable of complex interaction, as if they had a mind.[236] Environmental awareness and intentionality also appear to be present in unicellular and multicellular primitive species.[237] Thus, the amoeba lives as a unicellular being most of its life. When it needs food, however, it transmits signals to other amoebae nearby. Thousands of individual amoebae then come together and morph into a much larger entity that has new capabilities, such as the ability to move on the forest floor. Members of this larger entity release spores from which new amoebae are formed when they reach a better feeding area.[238]

As suggested by author Duane Elgin,[239] there may be a spectrum of consciousness that encompasses several levels, for example the atomic, cellular, and human levels. All levels would show a certain degree of awareness and intentionality. These abilities would make it possible to perceive but also to manipulate the environment. Mental processes and the *qualia* would be more refined in more complex beings.

Another central aspect of the postmaterialist paradigm is the deep interconnection that exists between the mental world (*psyche*) and the physical world (*physis*). Our usual perception of reality gives us the impression that *psyche* and *physis* are separated, but psi phenomena demonstrate that they are intimately linked. As Carl Gustav Jung and Wolfgang Pauli, who spoke about the *unus mundus* (the "One World"), I firmly believe that *psyche* and *physis* are closely related because they originate from a common principle: they are complementary aspects and manifestations of an indivisible whole that we can refer to as "Reality." For me, this first principle is a transcendental Consciousness that is also immanent, that is, ubiquitous in nature. Transcending energy, matter, space-time, and the world of the psyche, this Consciousness guides the unfolding of the physical and nonphysical worlds.

It is this indivisibility of *psyche* and *physis* that enables the mind to influence physical and biological phenomena. The remote mental interaction with physical systems or living organisms suggests that the unconscious part of the mind communicates – exchanges mental information capable of triggering action – with the physical world. On the other hand, telepathic experiences indicate that there are mental connections, usually unconscious, among human beings. The deep interconnection between *psyche* and *physis* does not seem to be based on quantum entanglement.

Indeed, non-local connections between entangled particles do not imply transfer of information[240] while mental interactions between humans, or remote mental influence on physical systems and living organisms, seem to involve an exchange of information implicating mental phenomena that are not taken into account in QM.

In addition to allowing us to subjectively experience our internal environment (psychic) and our external environment (located outside our physical body), the mind acts as a force – it has the ability to affect the state of the physical world. This observation is another key element in the postmaterialist paradigm. As we have seen, the mind influences the activity of the brain and physiological systems connected to the nervous system, as well as the activity of the genes. I have hypothesized that this psychosomatic influence is implemented neurobiologically by a process of psychoneural transduction: in this context, a transduction is a transformation of one form of information into another.[241] According to this hypothesis, mental processes (conscious and unconscious) are translated into neuronal processes, at various levels of brain organization (for example, molecular, chemical, and electrical). In a metaphorical way, we can say that the *mentalese* (the language of the mind) is translated into *neuronese* (the language of the brain). Thus, negative thoughts increase the secretion of cortisol while pleasant thoughts elevate the secretion of endorphins. The resulting neural processes are then translated into processes/ events in other physiological systems, such as the immune system or the endocrine system.

As I have shown in this book, in addition to its psychosomatic influence, the mind also has a non-local effect, beyond the limits of the body. In both cases (psychosomatic action and non-local influence), information, which does not depend on time or space, and information transduction, seem to play an essential role. On this subject, physicist David Bohm posited that there is a kind of active information in nature that makes the connection between the mental world and the physical world possible. The American physicist John Wheeler also believed that the universe was created by information.

As part of the *Theory of Psychelementarity*, which I proposed a few years ago,[242] I posited that the mind's ability to act as a force, whether somatic or exosomatic (non-local), varies among individuals. In order to be able to exert a greater force, the mind must be focused as a laser beam, rather than as a diffuse light. Rigorous mental training heightens this ability. Hence, some meditation techniques aimed at developing intentionally sustained concentration with respect to a specific object. According to the Theory of Psychelementarity, such training allows the mind to produce a greater mental strength.

If the mind is essential and cannot be reduced to matter, so what is the role of the brain? According to the postmaterialist paradigm, this organ acts as a filter and an interface for the mind. The hypothesis that the brain acts as a filter (or a "reduction valve") is not recent because it was proposed more than a century ago by philosophers Henri Bergson and Ferdinand Schiller, and psychology researchers William James and Frederic Myers.[243] This hypothesis was later taken up by author Aldous Huxley. Accordingly, the brain would normally limit access to expanded states of consciousness, and from an evolutionary point of view, this function would promote biological survival. During transcendent experiences, this filter function would be disabled to varying degrees. On a subjective level, such a deactivation would make it possible to experience domains of reality that are not physical.

Materialists claim that the fact that mental functions are disturbed when the brain is affected proves that this organ produces the mind. This is inaccurate because, as William James correctly noted, the hypothesis that the brain acts as a "transmitter" of the mind is also compatible with this fact. Rich mental experiences as well as confirmed perceptions occurring in a state of clinical death strongly support this hypothesis.

In order to understand the nature of mind-brain relationships, we can use television as an analogy. The TV receiver receives electromagnetic signals from a television station: these signals constitute the program. These electromagnetic signals are converted by the TV receiver into images and sounds. When an electronic component which is inside the receiver breaks, this can result in distortion, or even the total loss of the pictures and/or sounds because the receiver's capacity to receive or decode the electromagnetic signals is disturbed. Does this mean that the program is created by the TV receiver? Absolutely not. Similarly, mental functions cannot be reduced to the activity of brain regions and circuits.

In the context of the notion of the brain as a filter and the analogy of the TV, transcendent experiences could be compared to the temporary reception of higher frequencies. As a result, our field of perception and experience widens: the invisible becomes visible, and new aspects of reality are revealed to us.

## The Next Great Scientific Revolution

The implications of the postmaterialist paradigm are numerous and crucial. Thus, this new theoretical framework fundamentally changes the vision we have of ourselves, and it gives us back dignity and power as human beings by inviting us to develop the various aspects of our potential. The postmaterialist paradigm also encourages positive values such as respect, compassion, cooperation, and peace. In addition, by emphasizing the intimate connection between ourselves and nature, this paradigm encourages environmental awareness and the preservation of our biosphere.

The history of science has been punctuated by some special moments that have led to major paradigm shifts. Let's consider, for example, the transition from flat Earth theory to the conception of spherical Earth. Or that of geocentrism to heliocentrism.[244] It seems that we are now approaching another crucial paradigm shift – the transition from materialist science to postmaterialist science.

Far be it from me to play the prophet. However, I am deeply convinced that this transition will lead us to the next great scientific revolution, and that it will be even more vital for the evolution of human civilization than was the Copernican revolution. This revolution, which will affect our scientific understanding of the mind and its capabilities, will lead to a profound transformation of all aspects of life on Gaia – our wonderful planet, blue pearl floating in the immensity of space.

# Acknowledgements

I am very grateful to Bernardo Kastrup, the brilliant philosopher-scientist – and Publisher of Iff Books (an imprint of John Hunt Publishing) – for having recognized the value of *Expanding Reality*. I want also to thank the members of the support team at John Hunt Publishing Ltd. for their assistance and efficiency.

In addition, I am grateful to the scientists who have accepted to write endorsements for this book; in particular, I am indebted to Gary Schwartz, who has written the inspiring Foreword of *Expanding Reality*. Last, although many of them have left us, I cannot help but think of all those visionary scientists who helped lead us to the threshold of the next great scientific revolution.

# References

1.  Professor of Psychology, Medicine, Neurology, Psychiatry and Surgery; Director, Laboratory for Advances in Consciousness and Health, University of Arizona.

2.  There are various definitions of consciousness and mind. In this book, I refer to "consciousness" as the mental faculty that makes it possible to apprehend subjectively external phenomena (for example, a car approaching in the street) and internal phenomena (for example, our thoughts and emotions, including consciousness of our own existence, called self-awareness). As far as the "mind" is concerned, this concept refers, here, to conscious and unconscious mental processes and events (e.g., perceptions, thoughts, memories, willpower and imagination). In this context, the mind is not a substance (an entity), but rather a plurality of interrelated mental functions.

3.  Burtt, EA. *The Metaphysical Foundations of Modern Science*. London: Routledge, 1949.

4.  In this book, the terms "material" and "physical" as well as "materialism" and "physicalism" are used interchangeably. Physicalism is the doctrine that all that exists is physical in nature, that is, nothing exists except so-called physical things.

5.  Heisenberg, Werner. *Physics and Philosophy: The Revolution in Modern Science*. New York: Harper and Row, 1976.

6.  This flamboyant term was coined by American philosopher Neal Grossman.

7.  Beauregard M., Schwartz GE, Miller L., Dossey L., Moreira-Almeida A., Schlitz M. *et al.* "Manifesto for a Postmaterialist Science." *EXPLORE: The Journal of Science & Healing* 10, 2014, pp. 272-274.

8.  http://www.huffingtonpost.com/dave-pruett/toward-a-postmaterialistic-science_b_5842730.html.

9.  Beauregard M., Schwartz GE, Dyer N., Woollacott M. *Expanding*

*Science: Visions of a Postmaterialist Paradigm.* Tucson: AAPS Press, in press.

10. Bennett MR, Hacker PMS. *Philosophical Foundations of Neuroscience.* New York: Blackwell Publishing, 2003.

11. Kallmes DF, Comstock BA, Heagerty PJ, Turner JA, Wilson DJ, Diamond TH *et al.* "A Randomized Trial of Vertebroplasty for Osteoporotic Spinal Fractures." *New England Journal of Medicine* 361, 2009, pp. 569-579.

12. Kirsch I., Sapirstein G. "Listening to Prozac but hearing placebo: A meta-analysis of antidepressant medication." *Prevention & Treatment* 1: 2a, 1998. Document online: www.apa org/prevention.

13. Siegel B. *Love, Medicine and Miracles.* New York: Harper Perennial, 1998.

14. http: //www.nytimes.com/2012/08/12/opinion/sunday/beware-the-nocebo-effect.html _r = 0?.

15. http://www.newyorker.com/online/blogs/elements/2013/04/the-nocebo-effect-how-we-worry-ourselves-sick.html.

16. Fuente-Fernandez R., Ruth TJ, Sossi V., Schulzer M., Calne DB, Stoessl AJ. "Expectation and dopamine release: mechanism of the placebo effect in Parkinson's disease." *Science* 293, 2001, pp. 1164-1166.

17. Enck P., Benedetti F., Schedlowski M. "New insights into the placebo and nocebo responses." *Neuron* 59, 2008, pp. 195-206.

18. Beauregard M., Lévesque J., Bourgouin P. "Neural correlates of the conscious self-regulation of emotion." *Journal of Neuroscience* 21, 2001, RC165, pp. 1-6.

19. Lévesque J., Joanette Y., Paquette V., Mensour B., Beaudoin G., Leroux J.-M., Bourgouin P., Beauregard M. "Neural circuitry in voluntary self-regulation of sadness." *Biological Psychiatry* 53, 2003, pp. 502-510.

20. Perreau-Linck E., Beauregard M., Gravel P., Paquette V., Soucy J.-P., Diksic M., Benkelfat C. "In vivo measurements of brain trapping of $11^C$ $\alpha$-methyl-L-tryptophan during acute changes in mood states." *Journal of Psychiatry and Neuroscience* 32, 2007, pp.

430-434.

21. Davidson RJ, Lutz A. "Buddha's Brain: Neuroplasticity and Meditation." *IEEE Signal Processing Magazine* 25, 2008, pp. 176-174.

22. Fjorback LO. "Mindfulness and bodily distress." *Danish Medical Journal* 59, 2012, B4547.

23. Cahn BR, Polich J. "Meditation states and traits: EEG, ERP, and neuroimaging studies." *Psychological Bulletin* 132, 2006, pp. 180-211.

24. Travis F., Shear J. "Focused attention, open monitoring and automatic self-transcending: Categories to organize meditations from Vedic, Buddhist and Chinese traditions." *Consciousness and Cognition* 19, 2010, pp. 1110-1118.

25. Lutz A., Greischar LL, Rawlings NB, Ricard M., Davidson RJ. "Long-term meditators self-induce high-amplitude gamma synchrony during mental practice." *Proceedings of the National Academy of Sciences USA* 101, 2004, pp. 16369-16373.

26. *Ibid.*

27. Taylor V., Grant J., Daneault V., Scavone G., Breton E., Roffe-Vidal S., Courtemanche J., Lavarenne A., Beauregard M. "Impact of mindfulness on the neural responses to emotional pictures in experienced and beginning meditators." *NeuroImage* 57, 2011, pp. 1524-1533.

28. Hölzel BK, Carmody J., Vangel M., Congleton C., Yerramsetti SM, Gard T., Lazar SW. "Mindfulness practice leads to increases in regional brain gray matter density." *Psychiatry Research* 191, 2011, pp. 36-43.

29. Tang Y., Lu Q., Geng X., Stein EA, Yang Y., Posner MI. "Short-term meditation induces white matter changes in the anterior cingulated." *Proceedings of the National Academy of Sciences USA* 107, 2010, pp. 15649-15652.

30. Zachariae R. "Psychoneuroimmunology: A bio-psycho-social approach to health and disease." *Scandinavian Journal of Psychology* 50, 2009, pp. 645-651.

31. Vitetta L., Anton B., Cortizo F., Sali A. "Mind-body medicine: Stress and its impact on overall health and longevity." *Annals of the New York Academy of Sciences* 1057, 2005, pp. 492-505.

32. Ray O. "The revolutionary health science of psychoendoneuroimmunology." *Annals of the New York Academy of Sciences* 1032, 2004, pp. 35-51.

33. Cuijpers P., Smit F. "Excess mortality in depression: a meta-analysis of community studies." *Journal of Affective Disorders* 72, 2002, pp. 227-236.

34. Everson SA, Goldberg DE, Kaplan GA, Cohen RD, Pukkala E., Tuomilehto J., Salonen JT. "Hopelessness and risk of mortality and incidence of myocardial infarction and cancer." *Psychosomatic Medicine* 58, 1996, pp. 113-121.

35. Cohen S., Alper CM, Doyle WJ, Treanor JJ, Turner RB. "Positive emotional style predicts resistance to illness after experimental exposure to rhinovirus or influenza A virus." *Psychosomatic Medicine* 68, 2006, pp. 809-815.

36. Chida Y., Steptoe A. "Positive Psychological Well-Being and Mortality: Quantitative Review of Prospective Observational Studies." *Psychosomatic Medicine* 70, 2008, pp. 741-756.

37. Gordon JS. "Mind-Body Medicine and Cancer." *Hematology/ Oncology Clinics of North America* 22, pp. 683-708.

38. http://www.cnn.com/2011/HEALTH/03/03/ep.seidler.cancer. mind.body.

39. Lipton B. *The Biology of Belief*. Santa Rosa, CA: Mountain of Love/ Elite Books.

40. Church D. *The Genie in Your Genes*. Santa Rosa, CA: Energy Psychology Press, 2008.

41. Synonyms are "open" or "closed" as well as "expressed" or "repressed."

42. Lipton B. *The Biology of Belief*. Santa Rosa, CA: Mountain of Love/ Elite Books.

43. Dusek JA, Otu HH, Wohlhueter AL, Bhasin M., Zerbini LF, Joseph MG, Benson H., Libermann TA. "Genomic counter-stress changes

induced by the relaxation response." *PLoS One* 3, e2576.

44. http://www.bmedreport.com/archives/21586.

45. Jacobs TL, Epel ES, Lin J., Blackburn EH, Wolkowitz OM, Bridwell DA, Zanesco AP, Aichele SR, Sahdra BK, MacLean KA, King BG, Shaver PR, Rosenberg EL, Ferrer E., Wallace BA, Saron CD. "Intensive meditation training, immune cell telomerase activity, and psychological mediators." *Psychoneuroendocrinology* 36, 2011, pp. 664-681.

46. Mumford MD, Rose AM, Goshin DA. *An Evaluation of Remote Viewing: Research and Applications.* American Institutes for Research, 1995.

47. Schwartz SA. *The Alexandria Project.* Paris: Sand, 1985.

48. Mumford MD, Rose AM, Goshin DA. *An Evaluation of Remote Viewing: Research and Applications.* American Institutes for Research, 1995.

49. Jahn RG, Dunne B., Bradish G., Dobyns Y., Lettieri A., Nelson R., Mischo J., Boller E., Bösch H., Vaitl D., Houtkooper J., Walter B. "Mind/Machine Interaction Consortium: PortREG Replication Experiments." *Journal of Scientific Exploration* 14, 2000, pp. 499-555.

50. Radin D. *The Conscious Universe: The Scientific Truth of Psychic Phenomena.* New York: HarperEdge, 1997.

51. *Ibid.*

52. *Ibid.*

53. Honorton C., Schecter EI. "Ganzfeld target retrieval with an automated system: Model for initial ganzfeld success," in *RIP*, ed. Weiner DB and Nelson RB. Metuchen, NJ: Scarecrow Press, 1986, pp. 36-39.

54. Williams BJ. "Revisiting the Ganzfeld ESP debate: A basic review and assessment." *Journal of Scientific Exploration* 25, pp. 639-661.

55. Storm L., Tressoldi PE, Di Risio L., "Meta-analysis of free-response studies, 1992-2008: Assessing the noise reduction model in parapsychology," *Psychological Bulletin* 136, 2010, pp. 471-485. Utts J., Norris M., Suess E., Johnson W., "The strength of evidence

versus the power of belief: Are we all Bayesians?" in *Data and Context in Statistics Education: Towards an Evidence-Based Society*, Voorburg, The Netherlands: International Statistical Institute, ed. C. Reading, 2010.

56. Holt NJ. "Are artistic populations 'psi-conducive'? Testing the relationship between creativity and psi with an experiment-sampling protocol," in *Proceedings of the 50th Annual Convention of the Parapsychological Association*. Petaluma, CA: Parapsychological Association, 2007, pp. 31-47.

57. http://www.histoires-paranormales.fr/legendes-urbaines/churchill-premonition.

58. http://mystere-et-insolite.lo.gs/la-premonition-a28151992.

59. Dossey L. *The Science of Premonitions*. New York: Plume, 2010.

60. Honorton C., Ferrari DC. "Future telling: A meta-analysis of forced-choice precognition experiments, 1935-1987." *Journal of Parapsychology* 53, 1989, pp. 281-308.

61. Radin D. *The Conscious Universe: The Scientific Truth of Psychic Phenomena*. New York: HarperEdge, 1997.

62. Bierman DJ, Scholte HS. "Anomalous anticipatory brain activation preceding exposure of emotional and neutral pictures." *Toward a Science of Consciousness*, Tucson IV, 2002.

63. Mossbridge J., Tressoldi PE, Utts J. "Predictive physiological anticipation preceding seemingly unpredictable stimuli: a meta-analysis." *Frontiers in Psychology* 3, 2012, p. 390.

64. Bem D. "Feeling the future: experimental evidence for anomalous retroactive influences on cognition and affect." *Journal of Personality and Social Psychology* 100, 2011, pp. 407-425.

65. https://www.lesechos.fr/26/08/2016/LesEchosWeekEnd/00042-032-ECWE_et-si-vous-ecoutiez-votre-intuition-.htm#Azt8Q7BwUOWlegaV.99.

66. Guillemant P. *La Route du Temps*. Paris: Temps Présent, 2014.

67. Radin D. *The Conscious Universe: The Scientific Truth of Psychic Phenomena*. New York: HarperEdge, 1997.

68. Radin DI, Ferrari DC. "Effects of consciousness on the fall of dice:

A meta-analysis." *Journal of Scientific Exploration* 5, 1991, pp. 61-84.

69. Radin DI, Nelson RD. "Evidence for consciousness-related anomalies in random physical systems." *Foundations of Physics* 19, 1989, pp. 1499-1514.

70. Jahn RG, Dunne B., Bradish G., Dobyns Y., Lettieri A., Nelson R., Mischo J., Boller E., Bösch H., Vaitl D., Houtkooper J., Walter B. "Mind/machine interaction consortium: PortREG replication experiments." *Journal of Scientific Exploration* 14, 2000, pp. 499-555.

71. Jahn RG, Dunne BJ. *Margins of Reality*. New York: Jovanovich, 1987.

72. Radin D. *Supernormal: Science, Yoga, and the Evidence for Extraordinary Psychic Abilities*. New York: Deepak Chopra Books, 2013.

73. *Ibid.*

74. Radin D., Atwater FH. "Exploratory evidence for correlations between entrained mental coherence and random physical systems." *Journal of Scientific Exploration* 23, 2009, pp. 263-272.

75. Wahbeh H., Calabrese C., Zwickey H. "Binaural beat technology in humans: A pilot study to assess psychologic and physiologic effects." *Journal of Alternative and Complementary Medicine* 13, pp. 25-32.

76. Radin D., Atwater FH. "Exploratory evidence for correlations between entrained mental coherence and random physical systems." *Journal of Scientific Exploration* 23, 2009, pp. 263-272.

77. Radin D. *The Conscious Universe: The Scientific Truth of Psychic Phenomena*. New York: HarperEdge, 1997.

78. Emoto M. *The Hidden Messages in Water*. New York: Atria Books, 2005.

79. Radin D., Hayssen G., Emoto M., Kizu T. "Double-blind test of the effects of distant intention on water crystal formation." *Explore: The Journal of Science and Healing* 2, 2006, pp. 408-411.

80. Radin D., Lund N., Emoto M., Kizu T. "Effects of distant intention

on water crystal formation: A triple-blind replication." *Journal of Scientific Exploration* 22, 2008, pp. 481-493.

81. Radin D. *Supernormal: Science, Yoga, and the Evidence for Extraordinary Psychic Abilities.* New York: Deepak Chopra Books, 2013.

82. http://www.informationphilosopher.com/solutions/experiments/wave-function_collapse/.

83. https://en.wikipedia.org/wiki/Schr%C3%B6dinger%27s_cat.

84. https://futurism.com/john-wheelers-participatory-universe.

85. Radin DI, Michel L., Galdamez K., Wendland P., Rickenbach R., Delorme A. "Consciousness and the double-slit interference pattern: Six experiments." *Physics Essays* 25, pp. 157-171.

86. Radin D., Michel L., Delorme A. "Psychophysical modulation of fringe visibility in a distant double-slit optical system." *Physics Essays* 29, pp. 14-22.

87. Stapp H. *Mindful Universe: Quantum Mechanics and the Participating Observer.* Heidelberg: Springer, 2011.

88. Beck F., Eccles JC. "Quantum aspects of brain activity and the role of consciousness." *Proceedings of the National Academy USA* 89, 1992, pp. 11357-11361.

89. Penrose R., Hameroff S. "Consciousness in the Universe: Neuroscience, Quantum Space-Time Geometry and Orch OR Theory." *Journal of Cosmology* 14, 2011 – http://journalofcosmology.com/Consciousness160.html.

90. Grush R., Churchland P. "Gaps in Penrose's toiling." *Journal of Consciousness Studies* 2, 1995, pp. 10-29.

91. Tiller WA, Dibble WE, Kohane MJ. *Conscious Acts of Creation: The Emergence of a New Physics.* Walnut Creek, CA: Pavior Publishing, 2001.

92. Backster C. *Primary Perception.* White Rose Millennium, 2003.

93. Miller R. "The Positive Effect of Prayer on Plants." *Psychic*, 1972.

94. Backster C. *Primary Perception.* White Rose Millennium, 2003.

95. Tiller WA, Dibble WE, Kohane MJ. *Conscious Acts of Creation: The Emergence of a New Physics.* Walnut Creek, CA: Pavior Publishing,

2011.

96. http://www.metapsychique.org/La-Bio-PK-ou-l-influence-de-l. html.

97. Braud W. "Distant mental influence of the rate of hemolysis of human red blood cells." *Journal of the American Society for Psychical Research*, 1990, 84, pp. 1-24.

98. Watkins GK, Watkins A. "Possible PK Influence on the Resuscitation of Anesthetized Mice." *Journal of Parapsychology* 35: 257-272, 1971.

99. Radin D. *The Conscious Universe: The Scientific Truth of Psychic Phenomena*. New York: HarperEdge, 1997.

100. Braud W., Schlitz M. "Psychokinetic influence on electrodermal activity." *Journal of Parapsychology*, 1983, 47: 95-119.

101. Braud WG, Schlitz MJ. "Consciousness interactions with remote biological systems: Anomalous intentionality effects." *Subtle Energies*, 1991, 2, pp. 1-46.

102. Delanoy DL, Sah S. "Cognitive and physiological psi responses to remote positive and neutral emotional states," in Dick Bierman (dir.), *Proceedings of Presented Papers*, American Parapsychological Association, 37th Annual Convention, University of Amsterdam, 1994.

103. Achterberg J., Cooke K., Richards T., Standish LJ, Kozak L., Lake J. "Evidence for correlations between distant intentionality and brain function in recipients: a functional magnetic resonance imaging analysis." *Journal of Alternative & Complementary Medicine*, 2005, 11, pp. 965-971.

104. McCraty R., Shaffer F. "Heart rate variability: New perspectives on physiological mechanisms, assessment of self-regulatory capacity, and health risk." *Global Advances in Health and Medicine*, 2015, 4, pp. 46-61.

105. http://www.francetvinfo.fr/sante/biologie-genetique/l-adn-a-deux-nouvelles-lettres-qu-est-ce-que-ca-change_595171.html.

106. Rein G., McCraty R. "Structural changes in water and DNA associated with new physiologically measurable states." *Journal*

*of Scientific Exploration,* 1995, 8, pp. 438-439.

107. Rein G. "Effect of conscious intention on human DNA." *Proceeds of the International Forum on New Science,* Denver, Colorado, October 1996.

108. Sheldrake R. "Experiments on the sense of being stared at: the elimination of possible artefacts." *Journal of the Society for Psychical Research,* 2001, 65, pp. 122-137.

109. Colwell J., Schröder S., Sladen D. "The ability to detect unseen staring: A literature review and empirical tests." *British Journal of Psychology,* 2000, 91, pp. 71-85.

110. Schmidt S., Schneider R., Utts J., Walach H. "Distant intentionality and the feeling of being stared at: Two meta-analyses." *British Journal of Psychology,* 2004, 95, pp. 235-247.

111. Radin D., Schlitz MJ. "Gut feelings, intuition, and emotions: An exploratory study." *Journal of Alternative & Complementary Medicine,* 2005, 11, pp. 85-91.

112. Schmidt S., Schneider R., Utts J., Walach H. "Distant intentionality and the feeling of being stared at: Two meta-analyses." *British Journal of Psychology,* 2004, 95, pp. 235-247.

113. Gardner R. "Miracles of healing in Anglo-Celtic Northumbria as recorded by the Venerable Bede and his contemporaries: a reappraisal in the light of twentieth century experience." *British Medical Journal,* 1983, 287, pp. 1927-1933.

114. Sicher F., Targ E., Moore D., Smith H. "A randomized double-blind study of the effect of distant healing in a population with advanced AIDS. Report of a small-scale study." *The Western Journal of Medicine,* 1998, p. 169.

115. Radin D., Stone J., Levine E., Eskandarnejad S., Schlitz M., Kozak L., Mandel D., Hayssen G. "Compassionate intention as a therapeutic intervention by partners of cancer patients: Effects of distant intention on the patients' autonomic nervous system." *Explore: The Journal of Science and Healing,* 2008, 4, pp. 235-243.

116. *Ibid.*

117. Van Lommel P. *Consciousness Beyond Life: The Science of the Near-*

*Death Experience*. New York: HarperCollins, 2010.

118. Kellehear A. "Culture, biology and the near-death experience." *Journal of Nervous and Mental Disease*, 1993, 181, pp. 148-156.

119. Ring K., Elsaesser-Valarino E. *Lessons from the Light: What We Can Learn from the Near-Death Experience*. New York and London: Insight Books, Plenum, 1998.

120. Kellehear A. "Culture, biology and the near-death experience." *Journal of Nervous and Mental Disease* 1993, 181, pp. 148-156.

121. Ring K., Elsaesser-Valarino E. *Lessons from the Light: What We Can Learn from the Near-Death Experience*. New York and London: Insight Books, Plenum, 1998.

122. *Ibid.*

123. *Ibid.*

124. Van Lommel P. *Consciousness Beyond Life: The Science of the Near-Death Experience*. New York: HarperCollins, 2010.

125. Greyson B., Bush NE. "Distressing near-death experiences." *Psychiatry* 55, 1992, pp. 95-110.

126. Charbonier J.-J. *Les 7 bonnes raisons de croire à l'au-delà*. Paris: Guy Trédaniel éditeur, 2012.

127. *Ibid.*

128. Blanke O., Ortigue S., Landis T., Seeck M. "Stimulating illusory own-body perceptions." *Nature* 419, 2002, pp. 269-270.

129. Blanke O., Landis T., Spinelli L., Seeck M. "Out-of-body experience and autoscopy of neurological origin." *Brain* 127, 2004, pp. 243-258.

130. Van Lommel P., van Wees R., Meyers V., Elfferich I. "Near-death experience in survivors of cardiac arrest: a prospective study in the Netherlands." *Lancet*, 2001, 358, pp. 2039-2045.

131. Parnia S. *What Happens When We Die: A Groundbreaking Study into the Nature of Life and Death*. Carlsbad, CA: Hay House, 2006.

132. *Ibid.*

133. Clute HL, Levy WJ. "Electroencephalographic changes during brief cardiac arrest in humans." *Anesthesiology* 73, 1990, pp. 821-825.

134. Charbonier J.-J. *Les 7 bonnes raisons de croire à l'au-delà.* Paris: Guy Trédaniel éditeur, 2012.

135. Van Lommel P., van Wees R., Meyers V., Elfferich I. "Near-death experience in survivors of cardiac arrest: a prospective study in the Netherlands." *Lancet*, 2001, 358, pp. 2039-2045.

136. Lallier F., Velly G., Leon A. "Near-death experiences in survivors of cardiac arrest: a study of demographic, medical, pharmacological and psychological context." *Critical Care* 19 (Suppl. 1), 2015, p. 421.

137. Parnia S., Waller DG, Yeates R., Fenwick P. "A qualitative and quantitative study of the incidence, features and aetiology of near-death experiences in cardiac arrest survivors." *Resuscitation* 48, 2001, pp. 149-156.

138. Schwaninger J., Eisenberg PR, Schechtman KB, Weiss AN. "A prospective analysis of near-death experiments in cardiac arrest patients." *Journal of Near-Death Studies* 20, 2002, pp. 215-232.

139. The cerebral cortex designates the superficial layer of the brain, which is composed of grey matter.

140. Greyson B. "Cosmological implications of near-death experiences." *Journal of Cosmology* 14, 2011, pp. 4684-4696.

141. Blackmore SJ, Troscianko TS. "The Physiology of the Tunnel." *Journal of Near-Death Studies* 8, 1989, pp. 15-28.

142. Van Lommel P. *Consciousness Beyond Life: The Science of the Near-Death Experience.* New York: HarperCollins, 2010.

143. Parnia S. *What Happens When We Die: A Ground-Breaking Study into the Nature of Life and Death.* London: Hay House, 2008.

144. Whinnery JE. "Psychophysiologic correlates of unconsciousness and near-death experiences." *Journal of Near-Death Studies* 5, pp. 232-258, 1997.

145. Woerlee GM. "Cardiac arrest and near-death experiences." *Journal of Near-Death Studies* 22, pp. 235-249, 2004.

146. Van Lommel P., van Wees R., Meyers V., Elfferich I. "Near-death experience in survivors of cardiac arrest: a prospective study in the Netherlands." *Lancet* 358, 2001, pp. 2039-2045.

147. Greyson B. "Implications of Near-Death Experiences for a Postmaterialist Psychology." *Psychology of Religion and Spirituality* 2, 2010, pp. 37-45.

148. Jansen K. "Near death experience and the NMDA receptor." *British Medical Journal* 298, 1989, p. 1708.

149. Saavedra-Aguilar JC, Gomez-Jeria JS. "A neurobiological model for near-death experiences." *Journal of Near-Death Studies* 7, pp. 205-222.

150. Rodin E. "Comments on 'A Neurobiological Model for Near-Death Experiences'." *Journal of Near-Death Studies* 7, pp. 255-259.

151. Persinger MA. "Near-Death Experiences and Ecstasy: A Product of the Organization of the Human Brain." In S. Della Sala (ed.), *Mind Myths: Exploring Popular Assumptions about the Mind and Brain* (pp. 85-99). Chichester, UK: John Wiley, 1999.

152. Fenwick P., Lovelace H., Brayne S. "Comfort for the dying: five year retrospective and one year prospective studies of end of life experiences." *Archives of Gerontology and Geriatrics* 51, 2010, pp. 173-179.

153. Lawrence M., Repede E. "The incidence of deathbed communications and their impact on the dying process." *American Journal of Hospice & Palliative Care* 30, pp. 632-639.

154. Wills-Brandon C. *One Last Hug Before I Go: The Mystery and Meaning of Deathbed Visions*. Deerfield Beach, Florida: Health Communications Inc, 2000.

155. Greyson B. "Seeing Dead People Not Known to Have Died: 'Peak in Darien' Experiences." *Anthropology and Humanism* 35, pp. 159-171, 2010.

156. Charbonier J.-J. *Les 7 bonnes raisons de croire à l'au-delà*. Paris: Guy Trédaniel éditeur, 2012.

157. Moody R., Perry P. *Glimpses of Eternity: Sharing a Loved One's Passage From This Life to the Next*. Paradise Valley, AZ: Sakkara Productions, 2010.

158. Charbonier J.-J. *Les 7 bonnes raisons de croire à l'au-delà*. Paris: Guy Trédaniel éditeur, 2012.

159. http://www.inrees.com/articles/retour-vie-emi-guerison-spontanee/.

160. Gurney E., Myers F., Podmore F. *Phantasms of the Living*. London: Society for Psychical Research, 1886.

161. Greeley A., *The Sociology of the Paranormal: A Reconnaissance*, Beverly Hills, CA: Sage, 1975. Haraldsson E., Gudmundsdottir A., Ragnarsson A., Loftsson J., Jonsson S, "National survey of psychical experiences and attitudes towards the paranormal in Iceland," 1977, in JD Morris, WG Roll, RL Morris (eds.), *Research in Parapsychology*, Metuchen, NJ: Scarecrow Press, 1976, pp. 182-186.

162. True skeptics adopt an attitude of reserve and doubt in the face of any pseudo fact or proposition but keep an open mind. As for pseudo-skeptics, they adopt a negative attitude of denial, rather than an agnostic position, while continuing to call themselves "skeptics."

163. Dossey L. *Reinventing Medicine: Beyond Mind-Body to a New Era of Medicine*. New York: HarperOne, 2000.

164. http://www.adcrf.org/French/calm_stories.htm.

165. *Ibid.*

166. http://www.induced-adc.com/experiences/.

167. *Ibid.*

168. *Ibid.*

169. http://www.inrees.com/articles/Communiquer-avec-un-defunt-deuil-EMDR/.

170. *Ibid.*

171. *Ibid.*

172. http://www.windbridge.org.

173. Beischel J., Schwartz GE. "Anomalous information reception by research mediums demonstrated using a novel triple-blind protocol." *Explore: The Journal of Science & Healing* 3, 2007, pp. 23-27.

174. *Ibid.*

175. http://www.mondenouveau.fr/la-transcommunication-instrumentale-1-la-tci-audio.

176. http://www.slate.fr/story/100889/edison-voix-des-morts.

177. http://www.worlditc.org.

178. *Ibid.*

179. *Ibid.*

180. http://itcvoices.org/konstantin-raudive-and-his-itc-evp-breakthrough.

181. http://www.worlditc.org.

182. http://www.worlditc.org/h_22_estep.htm.

183. Estep SW. *Voices of Eternity.* New York: Fawcett, 1988.

184. http://www.worlditc.org.

185. *Ibid.*

186. *Ibid.*

187. Schwartz GE. "Possible Application of Silicon Photomultiplier Technology to Detect the Presence of Spirit and Intention: Three Proof-of-Concept Experiments." *Journal of Scientific Exploration* 6, 2010, pp. 166-171.

188. Charbonier J.-J. *Les 7 bonnes raisons de croire à l'au-delà.* Paris: Guy Trédaniel éditeur, 2012.

189. *Ibid.*

190. This is a philosophical reasoning principle which is also called principle of simplicity, principle of economy or principle of parsimony. Its name comes from Franciscan philosopher William of Occam (14th century). A modern formulation of this principle is that "The simplest sufficient hypotheses are the most likely."

191. Transduction is the conversion of one form of information/energy into another form of information/energy.

192. Keil HHJ, Tucker JB. "Children who claim to remember previous lives: Cases with written records made before the previous personality was identified." *Journal of Scientific Exploration* 19, 2005, pp. 91-101.

193. Tucker JB. *Life Before Life: A Scientific Investigation of Children's Memories of Previous Lives.* St. Martin's Griffin, 2008.

194. *Ibid.*

195. Stevenson I. *Reincarnation and Biology: A Contribution to the Etiology*

*of Birthmarks and Birth Defects*. Praeger, 1997.

196. Haraldsson E. "Psychological comparison between ordinary children and those who claim previous-life memories." *Journal of Scientific Exploration* 11, 1997, pp. 323-335.

197. Tucker JB. *Life Before Life: A Scientific Investigation of Children's Memories of Previous Lives*. St. Martin's Griffin, 2008.

198. Haraldsson, E. "Children who speak of past-life experiences: Is there a psychological explanation?" *Psychology and Psychotherapy: Theory Research and Practice* 76, 2003, pp. 55-67.

199. Tucker JB. *Life Before Life: A Scientific Investigation of Children's Memories of Previous Lives*. St. Martin's Press/Macmillan, 2015.

200. http://www.reincarnation.ch/regression-en-age/utilite.html.

201. Cheek DB. "Prenatal and perinatal imprints: Apparent prenatal consciousness as revealed by hypnosis." *Pre- and Peri-natal Psychology Journal* 11, 1996, pp. 97-110.

202. McCarty WA. *Welcoming Consciousness: Supporting Babies' Wholeness from the Beginning of Life*. Santa Barbara, CA: Wondrous Beginnings Publishing, 2004.

203. Dr. Michael Newton died in 2016.

204. Newton M. *Journey of Souls* illustrated, reprint. Lewellyn Publications, 2009.

205. *Ibid*.

206. *Ibid*.

207. Clark N. *Divine Moments*. Fairfield, IA: 1st World Publishing, 2012.

208. Hardy A. *The Spiritual Nature of Man*. Clarendon: Oxford, 1983.

209. Underhill E. *Mysticism: A Study in the Nature and Development of Man's Spiritual Consciousness*. New York: New American Library, 1974.

210. Stace WT, *Mysticism and Philosophy*, New York: Macmillan, 1960. Hood Ralph W., *Dimensions of Mystical Experiences: Empirical Studies and Psychological Link*, Rodopi, 2001.

211. Bucke RM. *Cosmic Consciousness*. New York: Dutton, 1969 (original publication: 1901).

212. http://www.issnoe.ch.
213. *Ibid.*
214. *Ibid.*
215. *Ibid.*
216. *Ibid.*
217. *Ibid.*
218. Grof S. *The Cosmic Game.* SUNY Press, 1998.
219. *Psychoactive Sacramentals: Essays on Entheogens and Religions.* San Francisco: Council on Spiritual Practices, ed. TB Roberts, 2001.
220. Kavenska V., Simonova H. "Ayahuasca Tourism: Participants in Shamanic Rituals and their Personality Styles, Motivation, Benefits and Risks." *Journal of Psychoactive Drugs* 47, 2015, pp. 351-359.
221. http://arutam.free.fr/Wakan.html.
222. Johnson MW, Richards WA, Griffiths RR. "Human hallucinogen research: guidelines for safety." *Journal of Psychopharmacology* 22, 2008, pp. 603-620.
223. Saver JL, Rabin J. "The neural substrates of religious experience." *Journal of Neuropsychiatry and Clinical Neuroscience* 9, 1997, pp. 498-510.
224. Beauregard M., O'Leary D., *The Spiritual Brain*, HarperCollins, 2007. Huxley A., *The Doors of Perception*, New York: Harper & Row, 1954.
225. Moreira-Almeida A. "Implications of spiritual experiences to the understanding of mind-brain relationship." *Asian Journal of Psychiatry* 6, 2013, pp. 585-589.
226. Weber SR, Pargament KI, "The role of religion and spirituality in mental health," *Current Opinion in Psychiatry* 27, 1994, pp. 358-363. Koenig HG, "Religion, Spirituality, and Health: A Review and Update," *Advances in Mind Body Medicine* 29, 2015, pp. 19-26.
227. Smith JC, Marsh JT, Brown WS. "Far-field recorded frequency-following responses: Evidence for the locus of brainstem sources." *Electroencephalography and Clinical Neurophysiology* 39, 1975, pp. 465-472.

228. Beauregard M., Paquette V. "Neural correlates of a mystical experience in Carmelite nuns." *Neuroscience Letters,* 2006, pp. 186-190.

229. *Ibid.*

230. https://hypnodio.com/language/en/home/.

231. To learn more about pseudo-skeptics: http://www. skepticalaboutskeptics.org.

232. Kuhn TS. *The Structure of Scientific Revolutions.* Chicago: University of Chicago Press, 1962.

233. Chalmers DJ. *The Conscious Mind: In Search of a Fundamental Theory.* New York: Oxford University Press, 1996.

234. Linde A. *Particle Physics and Inflationary Cosmology.* Chur, Switzerland: Harwood Academic Publishers, 1990.

235. Dyson F. *Infinite in All Directions.* New York: Harper & Row, 1988.

236. Cohen P. "Can Protein Spring into Life?" *New Scientist* 26, April 1999.

237. Baluška F., Mancuso S. "Deep evolutionary origins of neurobiology." *Communicative & Integrative Biology* 2, 2009, pp. 1-6.

238. Cohen P. "Can Protein Spring into Life?" *New Scientist* 26, April 1999.

239. Elgin D. *The Living Universe.* San Francisco: Berrett-Koehler, 2009.

240. Kafatos M., Nadeau R. *The Conscious Universe: Parts and Wholes in Physical Reality.* New York: Springer, 1999.

241. Beauregard M. "Mind does really matter: Evidence from neuroimaging studies of emotional self-regulation, psychotherapy, and placebo effect." *Progress in Neurobiology* 81, 2007, pp. 218-236.

242. Beauregard M. "The primordial psyche." *Journal of Consciousness Studies* 21, pp. 132-157.

243. Beauregard M. *Brain Wars.* New York: HarperCollins, 2012.

244. Schwartz GE. "Consciousness, Spirituality, and Postmaterialist Science: An Empirical and Experiential Approach," in LJ Miller (ed.), *The Oxford Handbook of Psychology and Spirituality.* New York: Oxford University Press, pp. 584-597.

# ACADEMIC AND SPECIALIST

Iff Books publishes non-fiction. It aims to work with authors and titles that augment our understanding of the human condition, society and civilisation, and the world or universe in which we live.
If you have enjoyed this book, why not tell other readers by posting a review on your preferred book site.
Recent bestsellers from Iff Books are:

## Why Materialism Is Baloney
How true skeptics know there is no death and fathom answers to life, the universe, and everything
Bernardo Kastrup
A hard-nosed, logical, and skeptic non-materialist metaphysics, according to which the body is in mind, not mind in the body.
Paperback: 978-1-78279-362-5 ebook: 978-1-78279-361-8

## The Fall
Steve Taylor
*The Fall* discusses human achievement versus the issues of war, patriarchy and social inequality.
Paperback: 978-1-78535-804-3 ebook: 978-1-78535-805-0

## Brief Peeks Beyond
Critical essays on metaphysics, neuroscience, free will, skepticism and culture
Bernardo Kastrup
An incisive, original, compelling alternative to current mainstream cultural views and assumptions.
Paperback: 978-1-78535-018-4 ebook: 978-1-78535-019-1

## Framespotting
Changing how you look at things changes how
you see them
Laurence & Alison Matthews
A punchy, upbeat guide to framespotting. Spot deceptions and
hidden assumptions; swap growth for growing up. See and be free.
Paperback: 978-1-78279-689-3 ebook: 978-1-78279-822-4

## Is There an Afterlife?
David Fontana
Is there an Afterlife? If so what is it like? How do Western ideas
of the afterlife compare with Eastern? David Fontana presents
the historical and contemporary evidence for survival of physical
death.
Paperback: 978-1-90381-690-5

## Nothing Matters
a book about nothing
Ronald Green
Thinking about Nothing opens the world to everything by
illuminating new angles to old problems and stimulating new
ways of thinking.
Paperback: 978-1-84694-707-0 ebook: 978-1-78099-016-3

## Panpsychism
The Philosophy of the Sensuous Cosmos
Peter Ells
Are free will and mind chimeras? This book, anti-materialistic
but respecting science, answers: No! Mind is foundational to all
existence.
Paperback: 978-1-84694-505-2 ebook: 978-1-78099-018-7

## Punk Science
Inside the Mind of God
Manjir Samanta-Laughton
Many have experienced unexplainable phenomena; God, psychic
abilities, extraordinary healing and angelic encounters. Can
cutting-edge science actually explain phenomena
previously thought of as 'paranormal'?
Paperback: 978-1-90504-793-2

## The Vagabond Spirit of Poetry
Edward Clarke
Spend time with the wisest poets of the modern age and of the
past, and let Edward Clarke remind you of the importance of
poetry in our industrialized world.
Paperback: 978-1-78279-370-0 ebook: 978-1-78279-369-4


Readers of ebooks can buy or view any of these bestsellers by
clicking on the live link in the title. Most titles are published in
paperback and as an ebook. Paperbacks are available in traditional
bookshops. Both print and ebook formats are available online.
Find more titles and sign up to our readers' newsletter at
http://www.johnhuntpublishing.com/non-fiction
Follow us on Facebook at
https://www.facebook.com/JHPNonFiction
and Twitter at https://twitter.com/JHPNonFiction

## *With Appreciation*

One does not fulfill one's success without the help of family and friends. I would like to acknowledge and thank the following:

First and utmost, my heartfelt thanks to my husband, Keith Hazelwood, and my sons, Joel and Jason Watkins, who continue to give me love and support. I love you!

My writer's group, The Wee Writers—Jan, Mary, Janet, Hallye, Ann, and Lilah. Their talent and friendship are such an inspiration to me.

My friends and former employees of my former business, Patches etc., who continue to cheer me on and occasionally share my travels.

Last, but certainly not least, is the AQS staff, especially Meredith Schroeder, who believed in this fiction series, and my patient editor, Elaine Brelsford, whose wisdom makes me a better writer. I feel they are on this journey with me and I hope to make them proud.

## *Dedication*

Christmas Traditions in St. Charles, Missouri have been in existence for over 40 years! The memories they have given visitors from across the country are priceless.

I am dedicating this book to the St. Charles Convention and Visitor's Bureau staff and the talented Christmas committee that works so hard each year to give us the best Christmas ever! Without my own enjoyment, I would not have been able to relate it to the *Christmas in Colebridge.* Thank you all so much, and continued success with the joy you bring to so many!

# CHAPTER 1

I thought of Moses parting the Red Sea as I walked through rows of hundreds of red poinsettias in my greenhouse. Kevin, my greenhouse manager, insisted we devote only one greenhouse to poinsettias for the first year. The other greenhouse stored a variety of plants and herbs for my flower shop, as well as my own house plants from 333 Lincoln. The poinsettias were in two sizes, and they were all my favorite colored variety, Red Velvet. The plan was successful from all indications so next year's crop would likely expand into another greenhouse with other colors and varieties.

"What do you think, boss lady?" Kip, my property manager, playfully asked.

"I'm certainly in my red element, Kip," I said with a glow. "This worked out beautifully! What do you think Beverly?" Her hands were on her hip assessing all the surrounding beauty as well.

1

"I was worried at first," she admitted. "They were all so tiny. I even had a dream that they all grew up to be different flowers than poinsettias, and Kip and Kevin were beside themselves!" We laughed at the thought.

I was pleased, for many reasons, that I had hired Beverly. She loved plants, of course, and her African-American grandmother had worked here at this historic Taylor House many years ago as a cook. Now her granddaughter was working for Dickson Properties, the new owners of 333 Lincoln. Sam, my deceased husband of over a year, and I had purchased and restored this historic home on the highest hill here in Colebridge. When Sam passed on, I took some of his generous inheritance and developed the Brody property next door which Sam and I had purchased right before he died. There was plenty of room for greenhouses which overlapped our residential property. The Brody barn became a storage facility for equipment, and the old Brody house served adequately for offices. It all happened quickly with a vision Sam and I both had at one time when we first bought the property after Mrs. Brody died. With only a nephew left, it had to be sold and Sam and I both knew she would have wanted us to own it.

"As soon as we move these poinsettias out of here Anne, we'll start on the geraniums for spring as planned," Kevin boasted. "We'll have to have more colors than red, I'm afraid!" They all snickered as they knew my favorite color was red in everything.

"The smaller poinsettias get saved for my open house Kevin," I reminded him. "I'd like to give one to each shop owner on Main Street, too! I'm sure there will be enough!"

"That's mighty nice, Miss Anne," Beverly commented.

"Aunt Julia and I are going to have our Christmas open houses at the same time since our shops are right next door to each other," I explained. "We'll both need plenty of poinsettias."

My Aunt Julia, who was divorced from my Uncle Jim, and her teenage daughter Sarah opened a paper gift shop called The Written Word. She had unusual paper supplies and a room dedicated to Jane Austen gifts, which my Jane Austen Literary Club enjoyed.

"Just tell me when you want them all delivered," Kevin said as he walked toward the door to leave.

"I will. Remind Abbey, your sweetheart, that she'll have her order for the Mistletoe Market delivered, as well, when she's ready." He grinned as he left.

Abbey was my employee from Brown's Botanical Flower Shop and she had an apartment on Main Street. She and Kevin had been dating about a year now, and they helped each other when they could. Abbey volunteered heavily for the Main Street Merchants and loved Christmas just as I did. She volunteered to chair the Mistletoe Market that would be set up in the park during this Christmas season. It was something new for the merchant's organization to make some money. Abbey often hinted that she would love to open her own Christmas shop on Main Street. She not only had a very creative eye as one of my designers, but a real feel for retail. She had moved here from New York so her ideas were fresh and unique to us in the Midwest.

"That market is going to be quite something, isn't it?" Beverly remarked. "I can see why everyone falls in love with charming Main Street and wants to have a shop there some day! I bet they don't realize how much work it is! Right so,

Anne?" I shook my head and smiled.

"No, I'm sure not, but it's a dream you see," I noted. "You never know what successes and failures life has in store for you. If someone would have told me that in my thirties I would be a widow with a nine-month-old little boy, I wouldn't have believed it." Beverly looked sad, feeling badly she brought up the topic.

"I don't know how you do it, Miss Anne," she said wrinkling her face. "I know Miss Ella is a big help, but Sammy's a handful. You have your shop and Dickson Properties to tend to also! I'm totally blown away just watching you!"

"I've always had good help, Beverly," I bragged. "When they say it takes a village to do something right, they're not kidding. I'm glad you are part of that Beverly. Your grandmother would be very proud." She looked down in embarrassment.

"You're right!" she claimed. "There are times I feel she's working right here beside me!" I laughed. "You may have already heard, but I have my Grandmother Davis' spirit here with me as well," I revealed. "For the most part, she behaves herself, but now and then she has to remind me that she's here! Someday I'll share that story with you and how she has been the mystery ghost up here on the hill since Marion and Albert once lived here." She looked confused.

"Oh yes, Miss Anne. I would like to hear all about that some time, but I bet Mr. Kip is wondering where I am about now," she said going toward the door. "I have to clock out shortly. Give little Sammy a hug for me, ya hear?"

I was hoping everyone who worked for me loved Christmas as much as I did. It was wonderful from a retail perspective, but even as a child, it seemed I waited all year for

Christmas to arrive. Main Street was the perfect place for the Christmas season. I knew most of the tourism budget was spent to enhance and promote this holiday because of its uniqueness. Visitors from all over the country knew about Christmas in Colebridge, and each year it seemed to get better and better.

# CHAPTER 2

When I went back in the house, Ella said she had just put Sammy down for a nap.

"He sure fights his naps and going down for the night, doesn't he?" Ella said as she fussed around the kitchen.

"I know Ella, he's just like his father," I reminisced. "He never had time to rest. I know he takes a lot of your time, Ella, when I can't be here. Can you handle it all okay? I want you to let me know because he is becoming a handful."

"I love that little fella like my own," she beamed. "I think it will get a little better when he can walk. There won't be as much lifting involved. He's getting so big."

"I wouldn't be so sure about it becoming easier, Ella." I quipped. "He's so quick on his feet. I suspect he'll be an early walker and a good size like his Father."

"Well, when that happens, he'll be running like his mama!" she nodded with a smile. "Speaking of mamas, don't

6

forget I'm going to your Mother's for dinner tonight. Are you here for the day, or will you be having to leave?"

"No, I'm here for the rest of the day," I assured her. "You really need a break so stay as long as you like tonight. I'm determined Sammy is going to continue that seven o'clock bedtime. I should have some time tonight to think about our Thanksgiving meal and who might be attending. It just seems like yesterday that everyone had gathered around the table here, doesn't it?"

"So true, Anne," Ella recalled. "Last Thanksgiving was one of the finest dinners ever! Did I hear you say Mrs. Dickson and her daughter Pat would not be coming this year? I bet they miss not seeing Sammy. He's her only grandchild, right?"

"Right," I said helping her wipe off the kitchen cabinet. "They have other plans, but they're determined to be here for Sammy's first Christmas!"

"I hope so," she added. "Don't you worry about how much longer Mrs. Dickson will be willing to travel? I also worry about your Mother and Harry. He really hasn't gained all his strength back since that last surgery."

"Don't you think it's because he has only one lung?" I wondered. "I think it's amazing that he functions like he does. Mother and he still take a walk each day, so that's good."

"Well, I'll be able to give you an update from them after my visit tonight," she reported with a wink.

Sammy awoke in his timely manner and he was all smiles. I could have piles of worry on my plate until I held Sammy in my arms. His disposition made it impossible to be mad at the world. His big smile with his two bottom teeth showing were filled with slobber that I devoured as I kissed him. I stayed in the nursery with him as he immediately squirmed out of my arms and onto the floor where he could go at his own speed and climb up on his feet. He was managing to go from one place to another holding on for dear life. I was so intrigued with this child's development that I found myself watching him in amazement. The nursery was carpeted and safer than some of the slippery hardwood floors in the house.

I pulled out one of many books that Mother kept supplying him from Pointer's Book Store where she used to work. She continued to write book reviews for their newsletter so she still frequented the shop. Sammy had his favorites and I happily obliged reading to him on the floor as if it were the first time. Mother reminded me that "Little Red Riding Hood" was my favorite story. I wonder if there was an edited version out these days that was minus the big bad wolf. Sammy loved when I included noises, as I did so dramatically when I read *The Three Little Pigs*.

He was saying "ma ma" upon my request. I knew most other children usually started out by saying "da da," an easier feat for sure, but there wasn't a da da anywhere around. He was attempting to say Ella, but it came out as "La La," which we thought was adorable of course. I took Ella in as our live-in housekeeper right before Sam died, and it had been working out beautifully.

We both waved goodbye to Ella as she went off to dinner. Of course there was a loaf of banana bread in her hand because she knew Harry loved it so much. She had been cleaning the house for Mother and Harry until Sammy came along. She found a replacement for them as she worked endlessly here at the Dickson house.

Tonight's dinner for Sammy included carrots which he disliked. I managed to cover them with peaches which he found tolerable. I researched to find the right organic, pure food for him, as did many new mothers these days. For the most part, he was a very good eater. Many nights I chose to eat a bite later in the evening after Sammy was in bed. There was always something left over from Ella's cooking, or I would just graze with a glass of wine. Everyone knew I wasn't a cook nor had any interest to learn. Sam loved to cook so we were the perfect match you might say. Sam had designed a custom kitchen that was envied by everyone when we started our restoration. He enjoyed it immensely, and now Ella was most appreciative.

When it was seven o'clock, I gave Sammy his bath and then listened to his usual fuss over being put in his crib for the evening. I went to my office in the room next door where I still had every intention of writing about the Taylor House. Now I had Sammy's journal to record in which I found to be much more fun to write about. Wouldn't it be grand, I thought, if Sammy would inherit my love for writing? When he settled down a bit, I concentrated on Thanksgiving plans on my paper pad. I was a list maker long before cell phones kept schedules and notations. I was also a visual person that had to picture in my mind any event I was planning. It was the quickest way for me to notice the details, which are so important.

As I did many a night, I went back into Sammy's nursery finding him asleep finally and covering him a bit. In his crib was the white-on-white lily quilt that, I assumed, came from the spirit of my Grandmother Davis. The quilt had appeared anonymously at my baby shower just as other gifts had on special occasions. She had a passion for lilies. I had this feeling of protection from the quilt. It was soft, and I truly felt it had magical powers from Grandmother.

For a good while, I stared at this lovely child who never got to see or meet his father. Finding out I was pregnant after Sam died was bittersweet. It had always been Sam's dream to have a family. I had put him off, which I now regretted. The stress of Sam's job at Martingale's gave him a massive heart attack, taking him at such a young age. I still had bottled up resentment for the company. As I continued staring at my precious son, I saw Sam's dark hair and smile. I admitted to having given Sammy my blue eyes. He was perfect, and he had many people who loved him. He was all I had that really mattered.

# CHAPTER 3

Sammy was an early riser like his father, except I used to awake smelling the aroma of Sam's coffee in the morning, not fussy chitter chatter. His happy mood was a good habit, and since he always slept through the night, I couldn't complain. His joyful manner always prepared me for a good day. After a diaper change we joined Ella downstairs in the kitchen. She couldn't wait to put her hands on him as she got him in his chair. I then went to change for my walk as she got his breakfast.

It was one of those cold, gray mornings with no sun in sight. The snow had melted so I managed to walk briskly down my hill to Lincoln Street. The baby weight had finally rolled off—to my delight. My active pregnancy kept things in control, and Sammy weighed only seven pounds and thirteen ounces when I delivered. I walked past Kip's office on the way. Sure enough, there was his truck, and smoke was coming out of the vent for the gas stove he had installed. I had lots to do so I decided not to stop and chat.

When I got back up the hill to the house, I helped myself to some hot coffee Ella had prepared. Little things like this luxury were so appreciated. Ella had become family when she cleaned our home for some time. She was in a financial straight. Since we had bedrooms to spare in this big house, it seemed like the right thing to do. It was a win-win for both of us. Little did I know at the time, she would also become a babysitter for my Sammy.

"So how was dinner last night?" I finally got around to asking her.

"The dinner was nothing fancy, but we had a good time," she recalled. "I brought home homemade bread from your Mother. We even played cards for an hour or so, then Harry called it a night."

"That's great!" I said looking around the kitchen for the bread. "So a slice of that bread sounds really good right now Ella! She isn't baking as much as she used to."

"Well, it's in that container, and I bet Mr. Sammy here would like a bite as well, wouldn't you?" she teased, kissing him on his plump cheek.

She was right! He smacked his lips with delight. Ella loved watching him eat so I was going to have to keep my eye on that or he'd be one chubby fella!

"You have book club tonight, right?" Ella reminded me.

"Oh, yeah," I responded. "You don't mind watching this little guy, I hope. Mother looks forward to me picking her up. I don't think she's getting out of the house as much with Harry's health being what it is."

"I noticed that too, Anne," Ella remarked. "She hates

leaving him alone, but I think she needs to get out more. He's the type to feel badly if he thinks she's giving up things because of him." I nodded.

"I better get changed for work," I said going toward the stairs. "I'll plan on going then."

Sally, my flower shop manager, was consulting early with a bridal appointment when I arrived at the shop. She treated the business as her own, and I truly appreciated her. I wished things would have worked out for her and Tim, but they didn't and life went on. I knew he had hurt her, but it was one of those relationships where she loved him much more than he did her. I saw it from the beginning. Some day she would find someone she deserved.

I was told David, my delivery guy, was out on a delivery to the Barrister Funeral Home. He had been such a good replacement after Kevin left. We missed Kevin's sense of humor, but David could not please us enough.

Jean and Abbey, the other two employees, were due in shortly. I quickly looked to see what orders had to be accomplished today before going to my office to check emails. It appeared to be a light day which was great for catching up. That meant making more floral arrangements for walk-ins and changing up displays in the shop for the Christmas season.

It seemed like my Uncle Jim's emails were almost coming daily. He had become over attentive after Sam died, thinking he had to watch over me and Sammy. He was Sam's best friend and, no doubt, felt an obligation. Since Uncle Jim and Aunt Julia's stressful divorce, things seemed to be much better between them. He was even helping Aunt Julia and their daughter Sarah with their shop. I was starting to feel a little crowded with his attention. However, I had always been an independent woman, which Sam knew very well, so for him to think I would depend on him, well, it was not going to happen. He would never be a father replacement for Sam. There was no doubt, however, that I would always be grateful to him for introducing me to Sam one Thanksgiving dinner at my house on Melrose Street.

Another email that consistently popped up was from my former boyfriend, Ted Collins. I'm sure he too thought his masculine attention would be helpful to me with Sam gone, but that also was not going to happen. I didn't want to continue to be rude to Ted, but my feelings were long gone. All I wanted to fill my life with right now was my beautiful son, my busy shop, and developing Dickson Properties. I had my hands full, and I felt Sam's presence more than ever.

"Might I interrupt, Miss Anne?" asked Jean with her English brogue. Jean had moved here from Bath, England, and she hadn't lost her accent and English ways. We loved it, of course, and I wanted to smile every time she spoke.

"Oh, sure, Jean, what's up?" I asked looking up.

"My Al asked if you would like him to spot a nice big tree for you again this year," she timidly asked.

"Oh my, yes," I answered. "The bigger the tree the better!" I grinned. "It's not too soon to keep an eye out for one," I said in agreement.

"Oh, and a jolly good one it will be!" she beamed. "You'll be at the club tonight then, Miss Anne? We have a lot to chat about with the Christmas Tea and all!" Jean had started a Jane Austen Literary Club in her home as she had such fondness for Jane and previously belonged to a club in England. Of course, most of us had joined and were enjoying it immensely.

"Yes, Mother and I both will attend!" I assured her.

# CHAPTER 4

O n the way to the Jane Austen club, Mother and I discussed plans for a simple Thanksgiving dinner. Our attendance would be quite small compared to other years. Sue, my cousin, and her two children would be going to see her parents who lived out of state. I'm sure my Uncle Ken and Aunt Joyce missed little Mia, who was now five, and Eli, just over a year. Sue had adopted the dark-haired darlings from Honduras, and they were quite loved and well cared for by many. Sue had a flexible job with the Barrister Funeral Home in their Grief Counseling department and was able to take advantage of the nursery care they provided for their employees. She filled her life with her children, but we wished that someday there would be someone just for her.

"So there will be just eight of us, plus Sammy, of course," I guessed. "I hope Amanda, Allen, and William will be able to join us for Christmas."

"I do too, Anne, but now they have extended families and we have to understand that," Mother said sadly.

"I do understand," I responded. "I just want them to know they're always welcome here."

Amanda and William were children from my Aunt Mary, Mother's half-sister. Aunt Mary was given up for adoption by my Grandmother Davis. We discovered the relationship in a quilt made by Marion Taylor containing cut up love letters in the back of the quilt. Albert Taylor was Grandmother's lover and the father of her child. Her resentment from the affair continues to haunt her here at 333 Lincoln, where the Taylors once lived. Amanda married Allen almost two years ago, and William now works out of state. They were thrilled to connect with our family which they didn't know existed, and we embraced them as well.

Jean's house was full of chatter and excitement when we arrived for the meeting. We were anxious to have that first cup of hot tea on this bitter cold night. Jean always had the best English breakfast tea which was our favorite. I saw a glass plate full of lemon bars that I knew had to come from Aunt Julia. They were always light and yummy.

"Ladies, ladies," Jean said clinking on the glass vase on the end table. "It's become a crowded night of sorts as we plan our Jane Austen Christmas Tea. Things are coming along quite nicely. Much to my delight, Donna can accommodate 140 guests at her tea room. We must firm up our menu which Miss Nancy has graciously planned. Miss Julia, our Queen of Etiquette, has taken her talents of The Written Word and designed our invitation of sorts, have you not, Miss Julia?"

"They really aren't invitations because these seats are going to fill overnight once the word is out," she bragged. "What I have done is taken card stock to create a welcome message to have at each place setting. It will be vintage cream in color and include a quote from Jane that says, 'Think only of the past, as its remembrance gives you pleasure.' Then at the bottom, it says something about making this Christmas Tea a pleasurable remembrance. On the backside, Jean has shared her endearing shortbread recipe." They all responded with approval and admiration.

## JEAN'S ENGLISH SHORTBREAD

2 sticks of softened butter; 2 cups of flour;
½ cup confectioner's sugar; and ¼ teaspoon of salt.
Preheat oven to 325 degrees. In a medium bowl, beat butter until creamy. In another bowl, mix flour, sugar and salt. Add dry mixture to creamed butter. Stir well, until the consistency of dough. Press dough into an ungreased 9X9 sized pan. Use a fork to prick the dough all over, about 20 times. Bake for 25 minutes or until done. While the shortbread is still warm, cut into squares. Enjoy with your favorite topping.

"It's quite lovely, Miss Julia, and most generous of you to provide such a hand-out," Jean added. "Other contributors are providing some grand attendance prizes such as gift tea baskets, Jane Austen pens, and the like. Anne is making holly centerpieces for the tables which will be smashing, I'm sure!" Everyone clapped in response. "You all should be delighted, should you not?" They all cheered once again.

Nancy now stood to get our attention.

"I'm working with Donna to make sure the delicacies are English and very appropriate, or it will be like any other tea," Nancy announced. "Donna said she'd like her servers to be in white aprons and caps from that period which would be a nice touch."

When reports on the tea were finished, I stood up to announce that everyone was invited to join the Christmas open houses that Aunt Julia and I were having after Thanksgiving. I told them we both had ordered some new Christmas items they may want to purchase for the gift giving season.

"Two for one. Quite jolly fun," Jean said as she clapped her hands. We chuckled at her accidental rhyme.

"Since we're making announcements, I want to tell you about something new that the Main Street Merchants are having this year," Abbey stated. "I am chairing a European Mistletoe Market we are setting up in a tent in our park along the river." Everyone stopped eating and drinking to hear more about this clever idea. "We hope to have lots of white lights, of course, to really light up the park. Besides selling Christmas items, candy, and cookies, we'll have live small Christmas trees, mistletoe, garlands, and wreaths for sale." Everyone responded with jovial questions and comments of approval.

"I have experienced such a market in my English hometown, and they are quite fun and magical if I must say!" Jean stated with excitement.

Christmas was already in the air. Questions and positive comments were coming from every direction. From all indications, it was going to be the best Christmas ever!

# CHAPTER 5

I was insistent upon inviting Sally to our Thanksgiving dinner. She had no other place to go even though she acted like she would prefer to be alone. She was ambitious in putting in extra hours at the shop as well as staying late for last-minute holiday sales and pick-ups. I finally convinced her she had to eat on that day, and she would hurt my feelings, as well as Sammy's, if she didn't join us, so she reluctantly agreed.

Abbey was going to join Kevin's family which was becoming a permanent picture. I would not be at all surprised to find an engagement announcement from the two of them around the holidays.

Ella and I firmed up a simple menu with an average-sized turkey that wouldn't have to be in the oven at the crack of dawn. I insisted Mother not bring one thing. I said if she happened to show up in the kitchen at gravy making time, I would be happy for her to take over as she did each year.

Aunt Julia shined at her baking skills so she was bringing a pumpkin pie and an apple pie, which I knew to be my Uncle Jim's favorite. He was always grateful so I kept including him in family affairs after their divorce. Aunt Julia told me Sarah wanted to bring her latest boyfriend to the dinner, but she and Uncle Jim both said she could not do so.

I had plenty of flowers from my own gardens to decorate with this year. We didn't have the first frost of the season as yet, so the mums Kip had planted around our house were in abundance in every color. I simply made a few arrangements for different tables throughout the house as well as an elongated one for our dinner table. It was fun and rewarding for me to do this for my home. Now that we had the greenhouses on Dickson Properties established, I wanted to do more floral gardening around the grounds and create special gardens.

I had no idea what I was going to wear for our dinner, but I had purchased a darling sailor suit for Sammy from the Baby La La shop on Main Street. I couldn't resist it when I saw it in the window. I spotted a red velvet jumpsuit for Christmas Eve as well so would likely go back to get it. He was growing so fast, I didn't know how to gauge sizes. I was so new to motherhood in general, it scared me sometimes. I caught myself whispering for Sam's guidance on occasion asking him what he would do. I suppose if I believed it helped me out, I saw no harm.

I took a quick peek in the poinsettia greenhouse to make sure Kip had designated all the ones going to the merchants on Main Street and the ones going to Abbey's Mistletoe Market on Black Friday morning. Main Street had to be decorated and lit by the day after Thanksgiving. Some merchants were decorating earlier and earlier each year trying to get the jump on the buying season. Since we were firmly told to use only live greenery on the outside of our buildings, the season was short.

A new feature besides the Mistletoe Market was a Christmas Quilt Show in the depot located in the park. They were only going to man it with volunteers on the weekends during December because they couldn't find enough help. Merchants were busy themselves this time of year so volunteering was an extra burden. The inside of the depot was spacious and rarely used making it the perfect place for the quilt show.

I didn't own a real Christmas quilt, but I told Isabella, owner of Isabella's Quilt Shop, that they were welcome to use my red and white Lone Star quilt which Helen gave me for my birthday one year. Isabella was a huge supporter of Main Street even though her quilt shop was not located there. Helen's daughter Pat, who occasionally won awards for her quilts, made it for Helen to give to me. I was touched that she acknowledged my love for red and white. I decorated one guest room in just those colors which I would once again arrange for Helen if she came for Christmas. I loved it! I'm sure she wouldn't mind the quilt being absent for the quilt show, and it might even be back home by then.

My personal present for me was always the big Christmas tree that would arrive a few days after Thanksgiving. Al and his friends did a great job of getting it in the door and setting it up straight. Sam helped the first year, but after that, it became too labor intensive. We had no ornaments the first year so we threw a party and asked everyone who attended to bring an ornament.

After that year, I shopped diligently in Door County, Wisconsin where I found a Christmas shop to answer all my needs. It was in a charming old church which was converted into a Christmas wonderland. We had a wonderful time together after Sam convinced me that I could really take time off to go on a trip. I brought back very long, personalized, appliquéd stockings for everyone at the shop including Sam and me. I recently ordered one special for Sammy, but it still had not arrived. I was thrilled it was still possible to do so. I wanted to quickly tell Sam about it, but as happened so often, I had to remind myself he was gone.

Life was happening fast, and it seemed new traditions were taking the place of old ones. The good news of it all was keeping my body and mind busy, away from any hurt.

# CHAPTER 6

The weather was gray with a grim forecast for Thanksgiving Day so I called Uncle Jim to see if he would mind picking up Harry and Mother. He was delighted, and said he would call them right away. I could tell he was thrilled to do so, and I had to remind myself not to take advantage of him.

Ella had an eye on Sammy as she prepared things in the kitchen. He was stacking the plastic containers she had given him; he loved stacking them and knocking them over. His hearty laugh and smile were infectious. While he was occupied, I set the dining room table and placed Sammy's chair nearby so he could be with us. I was hoping his afternoon nap would make for a pleasant mood at dinner time.

Before Sammy had his nap, I wanted to call and wish Helen a happy Thanksgiving. She loved hearing sounds and giggles from Sammy over the phone. If Sam were alive, he'd have us set up with her on Skype I'm sure. After a short conversation with her, Sue called to wish us the same from her parent's house. She said the trip was a challenge because of Eli having a cold. She wasn't sure she made the right decision to travel with him feeling cranky. She was hoping Mia wouldn't catch it next.

"Please tell Uncle Ken and Aunt Joyce hello and I hope we will see them at Christmas," I said with earnest.

"I will, but they are traveling less and less it seems just like Harry and your mother," she reminded me. "I really want my children to start having their own Christmas at their own house so I hope I can convince them to come."

"Well, just wait and see," I said trying to console her. "Give your sweet babies a kiss from me!"

Sue was trying so hard to create her own family. She, herself, was adopted and wanted to do the same since she wasn't married. She did resent that her parents didn't tell her about being adopted until she discovered it on her own. Now, if she could just find a nice Daddy for them. It now occurred to me that folks would start wishing the same for me which was not a pleasant thought.

Sure enough, light fluffy snowflakes were falling and covering the ground. While Sammy was still asleep, I showered and changed so I could be of more help to Ella. I made sure I was wearing some of the many pearls Sam had generously given me as I changed into one of my little black dresses with long sleeves.

Everyone started arriving early because of the weather. Sally had an SUV and was a brave driver, but she also arrived with the others. Ella was well prepared so drinks and appetizers were waiting in the living room. The aroma of roast turkey and all the goodies that go with it was well received by everyone. I brought Sailor Sammy down to meet and greet everyone, and he was in the perfect mood for all the attention.

When all were seated, I asked everyone to hold hands. "As the head of the Dickson household, I want to thank you for coming and ask that you join me in a prayer of thanksgiving." In complete silence, they took the hand next to them.

"Dear Lord, we are all gathered here with thankful hearts on this special day. We thank you for this wonderful meal and pray you will get our guests home safe and sound. We are especially grateful this year to have a new member of the Dickson family present at this table. Thank you for this special gift whom we all love and enjoy. Also bless those who are not with us this day, for they still remain in our hearts. In your name we pray, Amen." I wanted to break down in grief, thinking of Sam, but I had to be strong for all concerned. I saw Mother wiping away a tear so I looked away.

Smiles and laughter were heard around the table as everyone gave special attention to Sammy who was all smiles in his chair. His response, clapping his hands, said it all. He was happy to be here too.

To my delight, Sammy took a special liking to Ella's turkey dressing and of course his favorite mashed potatoes. After he had enough, Mother gave Sammy a new book that played music as each page was turned. It entertained him till we all went to the living room for dessert. Sarah was giving Sammy all her attention which he took full advantage of. She truly loved Sammy, and I think with a little training, she would make a fine babysitter one day.

Harry quickly finished his pie and seemed to be going to the window to check on the weather at every given moment. As he would report the snow coverings, moans and groans followed. I peeked myself and saw my lovely mums bending over due to the weight of the snow. Old man winter had arrived—ready or not.

After dessert and coffee, they all agreed to leave together. Uncle Jim thought it a good idea, in case any of them would need some assistance. They bundled up, gave hugs and kisses, and went home with full tummies.

# CHAPTER 7

Black Friday was here which meant all hands on deck for every retailer on Main Street and the whole town of Colebridge. The snow had stopped in the middle of the night so I was sure clearing would happen in time for the afternoon shoppers. Our hill was always cleared quickly, thanks to an arrangement Sam had made with some folks when we bought 333 Lincoln. I had heard the commotion in the early hours so I should be able to exit easily. I thought with some luck the snow would add to the Christmas kick off and put everyone in the spirit.

Ella knew this would be a long day so she suggested she send along some turkey sandwiches and pie from our dinner for my staff to nibble on through the day. I thought it a grand idea and, knowing Jean, she would be bringing something as well. I quickly gave Sammy a few mouthfuls of oatmeal and told him this was his very first snow day. I also told myself to write this and his first Thanksgiving in his journal.

"I want to bundle you up, my sweet, and take you out to touch the snow!" I teased. "We need a photo of your first snow! Why does this have to be a working day for your mummy?" He smiled as if he understood.

"Well, you can look at it that way, I suppose," Ella responded. "The good news is that he doesn't know what he's missing, and I just bet there will be many more snow days ahead in this little man's life!" I had to laugh as I agreed with her.

Ella scolded me for not getting on my way. I hugged and kissed my son, who was still in his pajamas as I grabbed the bag of leftovers from the kitchen counter to be on my way.

"I told Al straight away we were gonna get this blast of winter. Now they'll have to butcher the tree in the bloomin' snow," Jean fussed as I came in the door.

"Well, a good morning to you too, Miss Jean," I said taking off my coat. "Good morning, ladies. I'm sure the guys have cut trees in the snow before, Jean; I know Santa sure does every year!" They giggled.

"Right, Miss Anne," she grinned. "I surmise it will be a knockin' at your door in a night or two."

I was always intrigued on how Jean referred to nights rather than days.

"So will there be a party this year at the Dickson's?" Abbey asked while she put on her coat to leave for the Market.

"Yes, count on it," I said assuredly. "It's Sammy's first Christmas, and I want it to be very special."

"I'm sorry I have to leave you all, but I need to get down to the park to set up the Market for tonight's opening,"

Abbey announced putting her gloves on. "I don't know if I'll get back at all. If we don't get a turnout of volunteers to help, I won't be returning."

"Not to worry, Abbey, you have your hands full," I said to encourage her. "When David returns from a hospital delivery, he and Kevin will start delivering the poinsettias to all the shop owners and then some for your market. Don't worry, they'll get there in time. It will be magnificent!" She grinned still standing at the door.

"I think it will too, but you know we're already getting complaints from some of the shop owners that this market will compete with their business," she reported.

"Yes, I heard that from Phil, but don't worry about that," I said shaking my head. "You know it happens all the time from the same people. Frankly, I think it's going to add people to the area, and everyone will be just fine."

"Absolutely!" Sally added. "I bet Nick across the street is getting his knickers all riled up about this!"

We laughed, and I nodded.

"I already heard all about it! He's mostly upset because the market is selling Christmas cookies," I said.

"Oh my goodness, Anne, you should see how creative and beautiful they are," Abbey exclaimed. "The hospital auxiliary is making them from home for a fundraiser so we are sharing the profit with them. They're really delicious too. How could a Christmas Market not have Christmas cookies?"

"Oh, I must get some for Christmas!" I said with excitement. "That would be a big help to Ella, and Sammy will love them!"

"Better not tell Mr. Nick!" Jean teased. More laugher erupted.

"Be on your way, Abbey," I motioned with a wave. "We'll be fine. Break a leg!"

"What did you say, Miss Anne?" Jean asked in astonishment. Sally busted out with laughter.

"It's a good luck saying, Jean," I explained. She gave me a strange look not knowing whether to laugh or not.

# CHAPTER 8

I stayed open as late as the other retailers did last night and, there was no doubt, the Christmas In Colebridge opening was a huge success. The parade was so nostalgic, led by the fife and drum corp. Folks carried lit lanterns and sang Christmas carols as they walked down the whole length of Main Street. Joseph with Mary and the donkey were able to keep up as well. There were years when the donkey decided he wasn't moving, and that was that! I couldn't wait to check out all the shops and see what they had to offer for the season. I kept telling myself that one year I was going to take one of the carriage rides that pranced so elegantly down the brick streets. The horses had holly wreaths with bells around their necks, and the drivers wore Santa hats. Alongside each driver was a dog wearing a hat as well. The children loved the sight. It was a ride I would have enjoyed with Sam like so many other things I wished would have happened.

The leftover snow made the street picturesque and inviting. Aunt Julia stayed open later because she was located next to the costumed man roasting chestnuts on an open fire. I wondered if it turned out to be beneficial to her, or if she had any problem with the smoke that seemed to waft her way. It was a popular attraction that drew families experiencing it for the first time. I frankly didn't enjoy the taste, but obviously some did as they stood in a long line waiting for them to be roasted.

The next morning, my body ached for a nice long walk, but our hill was still a bit icy and I had a full day at the shop with Abbey's absence. We had a lot of unpacking to do, and we had to finish rearranging for our open house which was coming up. Feeling unrested, I played with Sammy at breakfast as he ate his Cheerios one by one. He had more energy than his mother did today, for sure. Eating one Cheerio at a time looked inviting to me.

"Oh, I forgot to tell you that a package arrived for you yesterday afternoon," Ella recalled. She went to the cabinet to retrieve it for me.

"Great, it's Sammy's Christmas stocking I ordered," I discovered. I unwrapped it quickly and, despite feeling there was a slight difference from the others I purchased, I thought it was great.

"Now what do you suppose Mr. Claus will put in such a big stocking?" Ella asked with sarcasm.

"I have no idea," I said laughing. "Sam and I just thought they were so fun when we saw them in Door County. Oh, Ella, I should have ordered one for you! Perhaps I still can, if I call them."

"Don't you dare, Miss Anne," she quickly responded. "I don't need a thing but some love and a roof over my head which I am lucky enough to have." She blushed.

"Yes, you do," I assured her with a big grin and a pat on the shoulder. "Let's go hang this up with the others."

I hung it next to Sam's stocking as if the three of us were still, indeed, a little family. We stood back to admire the display.

"I want Sam's stocking hung with ours every year, Ella," I said with urgency in my voice. "I want Sammy to be reminded of his father in every way I can." She smiled with approval.

I arrived a little later than I planned at the shop since I picked up some dry cleaning and then a favorite coffee of mine at Starbucks. Ella's coffee was a little weak, which she preferred, but I wasn't going to tell her it was not to my liking. I figure if she made it, she could have it the way she liked it.

"This telly's been ringin' up a storm for you, Miss Anne," Jean said immediately when I walked in tardy. "Here's a list of callers."

"Phil and Nick stopped by to personally thank you for the Poinsettias," Sally also reported. "I think you did a really fine thing there, Anne. They couldn't believe how generous you were!"

I went to check my emails and there were several more very nice thank you messages for the poinsettias. Why couldn't everyone on the street do more positive gestures like this instead of making life difficult for each other? Perhaps this would send a message. Hmmm.

Abbey came running in the door, almost out of breath. "Did you see the coverage of the opening last night you all?" Abbey asked holding the newspaper in her hand. She gave it to Sally first.

"Wow, is that Kevin kissing you underneath the mistletoe?" Sally teased, knowing the answer.

"Yes, but the reporter set it up," she said blushing. "We didn't think he was serious about using it."

As they laughed, I walked over to the paper to see it for myself.

"It is a Mistletoe Market for heaven's sake," I said with my approval. "This is great, and the other pictures are too! How were sales last night?"

"I was shocked!" Abbey said boldly. "I didn't know what to expect, but we sold out of the Christmas cookies two hours before we closed so I called them to bring lots more. They were thrilled."

"That's right nice because, after that, they went straight away to Mr. Nick's for cookies instead," Jean boasted. She had a good point.

"I'm off to the Market again," joked Abbey ready to leave again. "I hope I'm not leaving you all shorthanded here."

"We're fine," I assured her. "We've mostly been unpacking and rearranging for the open house."

"I'll be sure to be here!" she announced.

"Don't go breaking another leg, Miss Abbey!" Jean teased as she left. We all laughed, loving our Jean.

# CHAPTER 9

Late that afternoon Sue called to tell me she had returned home. Eli was much better, although she was concerned that his immune system continued to be so weak. It seemed like he was dealing with an ear infection or some other discomfort on a regular basis.

"Could you have lunch tomorrow, Anne, so we can catch up?" Sue pleaded. "I feel so left out after having been gone for your Thanksgiving dinner."

"I think it can work," I thought. "How about we meet around noon at Charley's to save time?"

"Great, see you then," she agreed. Charley's was down the street from my shop and was always the easiest place for many quick meals and gatherings. Sam and I frequented there on a weekly basis it seemed.

A slight knock on my office door, which was half open, was Jean telling me that Al had just called to say the

Christmas tree was cut and would be delivered tomorrow afternoon sometime. That was exciting news to hear, but I was going to have to prepare the entry way in the morning before having lunch with Sue.

"Did it all go okay?" I curiously asked.

"Jolly good from what I hear!" Jean answered with excitement. "I'll tell him straight away that you are in tune to the afternoon then." I nodded with approval.

I heard Isabella's voice out front so I out went to say hello. She was picking out two different Christmas arrangements.

"I was going to wait until your open house to get these but decided I needed them sooner," Isabella explained putting them on the counter to purchase.

"Thanks so much," I said as I started to wrap them in tissue for the outdoors. "I especially love the pinecone one. I think you're becoming quite the street person, Isabella." I chuckled. "I think you ought to seriously think about moving your quilt shop to Main Street."

"I've thought about it more than once, Anne, but the buildings along the street really aren't big enough for what I would need," she explained. "I've always been open to it especially when my rent keeps inching up."

"You do have quite a nice set up where you are," I admitted.

"That's what I've come to realize, at least for now," she noted. "Say, are you still willing for us to display your Lone Star quilt for the depot show? We'll be hanging the quilts this week in time for the weekend. We have plenty of quilts if you decide not to. I just hope we have enough volunteers to staff it for security. As you know, I'm a little paranoid about that." She was referring to the mystery of the Jane Austen Quilt that disappeared when she sent it away for competition.

"I'd really like to show it I think," I responded. "You're charging some kind of admission I hope."

"We really can't do that under our type of organization, but we are asking for donations when they leave the show," she explained. "I think we'll do fine. The rent was free on the depot so there's very little expense involved."

"I'll drop off the quilt at your shop sometime tomorrow then," I said as I tried to think about my schedule.

"Oh, that would be great. I want you to see the Santa Wall Quilt challenge that the Colebridge Quilt Club has displayed," she revealed. "The quilts are based on the International Santa characters from here on Main Street. They're quite good, and if I knew your sister-in-law Pat was coming, I'd have her judge them."

"I still don't have word on that," I shared. "She would love it, and she is so qualified. Helen cannot come without her, that's for sure. She hated missing Thanksgiving, and she's most anxious to see that grandson of hers as you can imagine."

"My, yes," Isabella responded. "I will find someone, but let me know if you hear, okay?"

I thanked Isabella once again as she went out the door. She was such a great supporter of Main Street as well as the whole community of Colebridge. Her husband died some years ago, and she poured herself into work. That's what people do to survive, as I was finding out.

When I returned home that evening, Ella was filling Sammy's sippy cup with water as he hung onto her skirt tail waiting for it. It was a Kodak moment that was pretty darn cute. He lit up when he saw me so I picked him up to give him an extra tight squeeze.

"Anne, I had a little scare today. I think everything is fine, but I wanted to mention that Sammy took a little tumble this afternoon," she revealed with a frightened look on her face. "He crawled up a few stairs, and by the time I saw him, he turned around like he was going to sit on a flat surface and tumbled down to the bottom."

"Oh, no! Is he alright?" I said trying not to panic.

"He cried, of course, because it scared him, but I looked him over really well and he seems fine," she anxiously explained. "He didn't hit his head that I saw. I didn't know whether I should call you or not."

"Oh, Sammy, what are we going to do with you?" I asked kissing him on his head. "You're right, Ella, he seems fine, but we're probably going to have to do more child proofing around here."

"I was thinking the same thing," she conceded. "He wants to go from place to place when he gets on his feet. I think he'll be walking before his first birthday, Anne."

"Well, we know how advanced he is," I teased. "We'll have to make some safe places for him in this big house."

The rest of the evening, I took Sammy upstairs for his bath and then read to him for a while before his bedtime. His attention span was short, something my Mother always said about me. Ella was right in that he was probably going to hit the floor running each day. Hmmm, who did that sound like?

# CHAPTER 10

The sun was bright and shining early the next morning which I thought was always a blessing in the winter. It was exciting to think about the arrival of the Christmas tree today. Rather than turn over to take advantage of Sammy's late sleep in, I got up to prepare the entryway for the tree. I went quietly down the stairs in my robe to get the paper and then put on some coffee. I would need Ella's help to move the round, hall entry table, but the other things I could move by myself as well as bring in all the boxes of Christmas decorations.

As I diligently went to work, waiting for the coffee to be finished, I thought about the years that Sammy would be adding ornaments to the tree. I had already purchased a couple of "baby's first Christmas" ornaments. I would also have Sarah make us some personalized ornaments when she demonstrated at their Christmas open house. I wanted one for Mother and Helen, too. Sarah had wonderful handwriting, was creative in many ways, and a good addition to The Written Word shop that she and Aunt Julia shared.

I liked to post photos now and then of Sammy in my shop's e-blast so Sammy helping with the big tree would be pretty cute. I could hear him chatter as he and Ella came down the stairs to join me.

"Well, it looks like Santa came here early, Sammy," Ella teased when she saw all the boxes. She handed him to me, but his frequent words of "down, down" were said as he wanted to squirm out of my arms and explore the boxes on his hands and knees.

"Not until you've had some breakfast, young man," I said as I put him in his chair.

My cell phone rang, and it was Uncle Jim. What could he want at this hour?

"I have a vacation day today, Anne, so thought I'd check to see if you needed help with anything," he began. "It's going to be a nice day, and thought I could even take Sammy out for a walk in his stroller."

That was odd, I thought.

"Oh, thanks, Uncle Jim, but they're bringing the tree here this afternoon so I need to be here." I explained.

"Well, then I can probably help with the tree or take Sammy so he's out of the way," he insisted.

I was uncomfortable with his assertiveness, and I didn't ever want Sammy out of the way.

"You're welcome to stop by if you like, but I have plenty of help," I said with little encouragement.

"Alright then," he said with disappointment in his voice.

After everything was clear for the tree, I showered and dressed to drop off my quilt at Isabella's and then meet up with Sue for lunch.

"These are marvelous, Isabella," I remarked when I saw the Santa quilts all displayed. "Did you do one of these, Norma?" I knew Norma, her employee, was an excellent quilter. She had come to Mother's to help quilt Isabella's quilt one afternoon.

"Oh, no ma'am. I couldn't do that while I'm still working here," she explained. "This is strictly for our customers and very good ones we have, I might add."

"Do you have a favorite, Isabella?" I asked while I was getting a closer look.

"I don't dare say, but yes, I do," she confessed. "I would get in lots of trouble stating my preference."

"There should be pictures and an article in the newspaper about these!" I bragged. "I'll tell Kathy, whose doing the publicity for Christmas In Colebridge. Maybe she can do something. You always have great ideas Isabella! I'm glad I got to see these. I need to go now. Thanks for taking my quilt!"

Sue was already at the counter waiting for me at Charley's when I arrived.

"How about we eat here at the counter?" Sue suggested. "The service is always faster."

"Good idea," I said. "I'm starving. I think one of their hot and juicy hamburgers on this nippy cold day sounds good."

"I have to have their French dip like always," Sue stated. "It's a habit, but they're so delicious."

It was Sam's favorite here, too, but I didn't say anything. It was little things like this that still gave me a sickening feeling. I wasn't sure how much and how long I could keep talking about Sam. I would talk about him constantly if I could. I didn't know what was normal.

Sue pulled out a brochure and began to explain what it was all about.

"This was sent to the office and passed around, but it sounds so neat to go to," she started to explain. "It's called the German Taundernbaum Church Tour. There are these little towns southeast of here that decorate their little country churches with live Christmas trees and other decorations. They also serve refreshments at each place, like German coffee cake, made by the church ladies. Some have live music which has to be so neat. Mrs. Schmidt, at work, said it's the best Christmas experience she has ever had. She said she knew one tiny church was heated with just a wood stove."

The whole event sounded wonderful.

"I thought at first my children would enjoy it, but it would actually be a nightmare when I think about taking them. I was thinking it would be nice if you, you're mother, Aunt Julia, and I could all go together. I'll take my van. Driving through all those winding roads is beautiful, but if the weather is bad, we shouldn't be going."

"Sure, I know that area," I said remembering. "Now, I would love to go, but it sounds like something Ella would really enjoy going to as well. She doesn't get out much. I would just have to find someone to take care of Sammy."

"I thought of that, too," Sue said smiling. "My little guys will be at the Barrister Nursery, and I know Nancy would be more than glad to let you have Sammy stay there. They have wonderful sitters. You know you need to get Sammy around some other children, Anne."

"Yes, I know," I admitted. "He's around all those old people." She laughed. "I think it sounds great. If you can get Nancy to agree to take him there, I'll go!"

"Wonderful!" Sue said as they were bringing our food.

# CHAPTER 11

Ella and I were ready for Al and the tree crew. The Persian rug was rolled up and the furniture removed. We laid sheets all over the hardwood floors to catch the pine needles and debris from the tree as well as their dirty shoes. I learned this the hard way from the first tree I had delivered. Ella made sandwiches and a tray of brownies for the workers as a treat. Sammy was still down for a nap when I heard the truck come up the hill.

Al was in charge of the crew and was enjoying every minute of it. I opened the front double doors and couldn't look as they squeezed in the huge, wrapped tree. Ella watched from the kitchen door with her mouth wide open in expectation. Once in the house, Al climbed the extra-tall ladder to hold the tree as the others raised it into its stand. I had never witnessed the step by step process before so it was fascinating to see it proceed. Once the tree was standing upright, the binding twine was cut, and the branches fell into place. As I took photos on my phone, the tree appeared to be taller than last year, but not as wide. I told Al to leave out the ladder for use in decorating.

"I'll be happy to stay and help you put up the lights, Anne," Al said as they tried to clean up.

"No, you all have done enough," I said with gratitude. "Please stay and have some refreshments will you?" I then remembered that Al was a recovering alcoholic and shouldn't have a beer like the other two guys. It nearly cost him his marriage with Jean. Luckily, I had other beverages to offer.

Before they left, they removed the sheets and helped us put the furniture back. Even the bare tree was absolutely beautiful, and the smell was heavenly! I was hoping Sam was watching with a big smile.

Ella went up to get Sammy, who was crying in his crib, as I saw the crew out the door. The south porch had needles everywhere, but it would have to be cleaned up at another time.

Sammy was fed his dinner, and then Ella and I started sorting out the miles of Christmas lights as Sammy looked on from his padded floor seat. There was a knock at the door when I got to the top of the ladder to begin attaching the string of lights. Ella answered, and it was Uncle Jim.

"Hey, how's that tree coming along?" he asked coming in to join us. "Looks like I came at just the right time."

"Yes, the hard part is done!" Ella said to Uncle Jim. "Isn't it a grand tree?"

"It sure is," he said looking up. "How about I give you a hand once I hug that cute little fella over there?" He picked Sammy up to hug and kiss as if he were his very own.

"How would you like a drink or something to eat, Jim?" Ella asked to be polite.

"Oh, no thanks, Ella, but let's see if I can help get this tree lit," he said picking up the string I was working with.

I said very little as he took charge of the nest of lights on the floor. I stayed on the ladder as he fed me the lights. I had to admit that it was a big help. Ella gave Sammy an empty box to play with so he was entertained. At last we seemed to have the tree covered in white lights. I stepped off the ladder and stepped back to admire our work. We turned off the room lights, and even Sammy's eyes were focused on the beautiful sight.

"Thanks, Uncle Jim, this was a big help," I said with appreciation.

"I've told you before, you don't have to keep calling me Uncle Jim," he responded, moving closer to my face.

"You'll always be my Uncle Jim," I said as nicely as I could.

"Have it your way then, Annie," he said grinning. Did he just call me Annie? No one called me Annie but Sam, ever!

"Don't ever call me Annie, Uncle Jim," I said without thinking twice. "That was what Sam called me." He knew he had struck a bad nerve.

"I'm sorry; I said it without thinking," he said looking down. "I'll be moving on. Say, I hear there's going to be a Christmas party here sometime soon. Sure hope I get an invite."

"You know you will," I said smiling to ease the tension. "You may have to help bartend!"

"Just let me know, I'll be happy to," he said as he went over to kiss Sammy goodbye.

There was no kiss on the cheek for me like Uncle Jim

always did. What was different? I couldn't put my finger on it.

"Time for bed, Sammy boy," I said taking Sammy out of Ella's arms. "We have to see what Billy the Bear is up to." It was a new book from Mother that Sammy seemed to enjoy the last time I read it to him.

The next morning I stopped by Mother's to see how she and Harry were doing. Her phone calls to me were becoming fewer and fewer. As always, they were sitting at the kitchen table, drinking coffee, and reading the morning newspaper. "How are you feeling, Harry?" I asked kissing him on the cheek.

"I wish I could say fit as a fiddle, but I'm slowing down a bit," he explained with frustration. "That mother of yours has to do almost everything by herself these days. I couldn't even help her bring up the Christmas tree from the basement yesterday."

"Why didn't you wait for me to help you with that, Mother?" I asked in concern. "Are you sure you still want to go to the trouble of having Christmas Eve dinner? You know I can do both over at my house."

"Don't talk like that, Anne," she scolded. "Of course Christmas Eve dinner will be here just like always. We're not dead yet, are we, Harry?"

*Ouch, I guess I went too far.*

"That's fine, I just don't want you to go to a lot of work," I explained. "Ella and I can certainly bring some things to be helpful. How about I come over one night next week and help you get that tree up like old times?" She grinned.

"Only if you bring that grandson of mine," she teased.

"I'll keep the little man occupied unless he's walking by

now," Harry teased. "I may not be able to keep up with him if he is."

"He's not walking yet, but he crawls faster than anyone I've ever seen," I reported. "He's entertained easily with books so perhaps that could be a suggestion for you."

"That would be wonderful, sweetie, if you have the time," Mother said affectionately.

I got in the car knowing we were starting a transition stage with the two of them. Harry was slipping faster with his health, and Mother's worry was starting to show on her.

# CHAPTER 12

When I arrived at the shop, I saw Aunt Julia's car parked out front. I decided to go in and confirm our plans for our open houses. She was in her office with her head in her hands like she had a bad headache.

"It's not that bad, is it?" I said, surprising her.

"Oh, Anne, these numbers drive me crazy!" she said in frustration.

"What do you mean?" I asked, seeing she was serious.

"It's probably the Christmas season, but the revenue isn't keeping up with the new purchases, or I should say Sarah's purchases. The bills just keep flying in."

"Don't you approve what she picks out?" I asked feeling concerned. I sat down next to her. This was not good news.

"I've been trying to support her taste and energy here in the business," she said shaking her head.

"Well, that goes along with being able to pay for it," I stated firmly. "She should be here with you now, helping to figure this out. She probably thinks it's like at home where you take care of everything. The easy part of retail is the buying and creating, but it takes money to do that. It also takes time to know how to time everything with deliveries so you'll be able to pay for them. Once your credit is affected, then you're really sunk."

"Oh, I'll get them paid. It just may not come from shop money," she admitted.

"No, no, don't start that!" I said boldly. She looked at me like I must have an explanation. "Business is business, and your personal life should be separate. Are you paying yourself a salary? It's none of my business, but that's the first thing my accountant told me to do. You'll never feel like you can afford to pay yourself if you don't start with something minor, just something! You do not have a hobby here. You are just not playing store, Aunt Julia. You have a business that's supposed to turn a profit." She snickered.

"Well, I just thought that the first year we would put everything back into the business since I don't need this to live off of," she revealed.

"That's very commendable and many shops do that, but you're not helping the health of the business by doing that. It has to sustain itself. I know what rent you are paying to Bill, and it isn't cheap. I really want you to succeed. You're being so generous with your giveaways, but it all adds up. You have to think about what the return might be. Okay, I've said enough, it's your business."

"I think our Christmas business will help, and I do appreciate what you've said, Anne," she said, rubbing the back of her neck. "I've already adjusted some things for the open house."

"Like cheap wine instead of champagne?" I teased. "That's why I'm here really. You agreed to the hours I sent you in an email?" She nodded. "Kathy is making sure we have carolers outside singing most of the time which should draw folks to this part of street."

"Wonderful, and thanks for all the advice," she said, getting up to give me a hug.

"I didn't mean to bring you down further," I said sadly. "You have to bring Sarah in on this, Aunt Julia. She may be more mature about this than you think." I walked out of her shop hoping I had done the right thing.

When I got back to my shop, I notice that Abbey was filling in for Sally, who had a dental appointment. I showed them the leftovers Ella had sent, and their eyes lit up especially David's. What was it about men's bottomless stomachs that never caused them to gain weight?

"I was just telling Jean and David that our Market had a thief in the night!" Abbey said with anger in her voice.

"You're kidding!" I quickly responded. "What happened?"

"Well, there's been some things moved around, making somewhat of a mess, but the worst is they've cut the Christmas tree lights twice now," she said. "They're sharply cut with wire cutters, Kevin thought. We want those trees lit all the time for security reasons, and it looks wonderful driving by, but apparently someone doesn't agree with us."

My mind immediately rushed to some of the merchants who complained about the Market competing with their shops, and then the party of no, who didn't want any lights whatsoever on our historic street. This Market had to be a nightmare for them.

"It's a reckless shame, by golly," said Jean in disgust.

I was speechless.

"Kevin thinks it's probably just some kids or drunks who are getting a kick out of the vandalism," Abbey shared. "He thinks he could put a camera somewhere to maybe catch who it is."

"Everyone always wants to blame kids," I said with anger. "Like all places and organizations, there are mean girls and mean boys who don't want to play nice. Have you called the police?"

"No, George thought we would get bad publicity and we should try some means ourselves to stop it," she explained. "It's costing the merchants plenty with buying more lights. I thought I could take turns with a few others and spend the night there to see who shows up."

"And freeze your tippy toes off!" Jean threatened. "The little buggers aren't worth that!" We had to laugh, and yet it was a valid statement.

"The place isn't heated," I reminded her. "I think you need to call the police so that you have it on record. They could make more observant visits at least when they drive by."

"They thought we could take turns going over to the depot to warm up," Abbey suggested. "We're already doing that with the workers. We have to figure out something quickly I'm afraid."

# CHAPTER 13

The precious events of December were starting to cram my calendar. I was making lists on top of lists, and each day was filled with preparation for the holidays ahead.

Nancy called to confirm that Sammy would be welcome at the Barrister Nursery anytime. She thought the church tour sounded wonderful and wished she could join us.

"I don't know why our churches here don't do something like that," Nancy noted. "Hey, I have a favor to ask you, if I might."

"Sure, what's up?" I asked with curiosity.

"Richard and I are chairing a wine tasting Fundraiser for an organization that provides a better Christmas for children of parents who are in prison," she explained. "It's a very worthy cause, and I was going to hit you up for a gift certificate for their silent auction, but there's really more of another reason."

"Well, the donation is no problem, but what else do you need?" I wanted to know.

"We want you to be our guest," she began. "It's next weekend and is going to be a very nice affair at Rascino's. They're closing the place to accommodate this event. Your social life has to be a bit more than Main Street and flowers, don't you think?"

"Oh, Nancy, I couldn't possibly do one more thing before Christmas," I said without hesitancy. "I am gone so much that any extra moments need to be spent with Sammy."

"That's very admirable, Anne, but I happen to know he's in bed by seven o'clock and you have an in-house babysitter so that excuse won't work," she rattled as if she were prepared for my response. "I want this to be a little Christmas gift from Richard and myself. We thought about what we could possibly give you and came up short. We did, however, think that you could use a grown-up night out. You invite us all the time. This time, it's our treat, okay? We'll be happy to pick you up if you like." It did sound pretty harmless and it was for a good cause. I paused.

"Okay, I'll try, but I'll drive myself so I can leave if I need to," I finally said. One glass of wine and I could escape without feeling badly.

"That's terrific, but you won't want to leave this fabulous affair," she promised. "Lot of important people from Colebridge are putting a lot of work into this to make it spectacular! I bet many of them are your customers, Mrs. Dickson!"

"Okay, okay, Miss Nancy Brewster." I called her by her maiden name when she drove me crazy. She always seemed to get her way with me.

Mother was expecting Sammy and me for an early dinner so I could help put up her Christmas tree. We were received with open arms and a pot full of wonderful smelling chili. Sammy didn't disappoint them with his cute antics and loving personality once he was there for a while. Ella always said Sammy was a mama's boy, but what else would he be? I was okay with him being a little cautious of folks. As soon as we finished eating, Mother insisted I take one of the pies home for Ella to enjoy. One was peach, and the other was banana cream.

"I'm taking one of these to a luncheon reception for our new pastor tomorrow so just pick which one you want," she revealed.

"What new pastor?" I quickly responded. "What happened to our Reverend Hamel? He was so wonderful at our wedding."

"He accepted a call from somewhere up north where his wife's family is from," she explained. "Now if you came to church more often, you would know all this my dear." She had me there!

"Don't let your mother razz you, Anne," said Harry jumping into the conversation. "Frankly, this new young man is quite impressive, and I think he'll be good for that congregation."

"He's a handsome fella, I'll give him that. Maybe you can go with us sometime through the holidays. Of course, I would like to show off that grandson of mine, too. We'll see in time how he does. I really do miss Rev. Hamel." She grinned at me.

"I think we can manage that," I said to make her happy. "Now, let's get these ornaments on the tree. I am so, so happy you bought this pre-lit tree a couple of years ago. Remember when we would squabble every single time just like you and Dad used to do?" She laughed. "Uncle Jim stopped by and helped me finish up that horrible lighting task on my big tree. I love my big, live tree, but perhaps someday, I'll have to go artificial. It's a lot of work. Beverly and I did all the ornaments, and she also did a lot of outdoor lighting at home for me. Wait till you see what all I have lit up this year, Mother!"

"Well, that's nice," Mother said. "What all did you light?"

"We now have lights on a couple of pines out front, lights around the gazebo, and even lights on the barn and office," I bragged. "I suspect we now light up the entire hill at 333 Lincoln." They laughed.

"We'll have to check it out, Harry!" she told him. He didn't hear her because he was busy trying to get Sammy to look at his book. It became harder to do so as he wanted more and more of my attention as we were getting into those pretty ornaments. Wanting out of Harry's lap was a cute way of paying me homage in that I would know what he needed. Being a Mom was pretty cool! We were nearly finished when his fussing became louder so Mother suggested that she could finish up at another time.

"Okay, Mr. Dickson, you succeeded," I said as I picked him up. "Let's get your coat on and head home to beddy-bye. It's past your bedtime." We kissed and hugged good-bye as Harry carried out the peach pie for Ella.

# CHAPTER 14

Brown's Botanical never had a Christmas Open House before. Today was the first for us. I never considered us a gift shop like so many other shops on Main Street. I knew they were very successful, so when Aunt Julia said she wanted to have one, I decided to give it a go. Sally was especially happy for us to try it and ordered merchandise that she thought people could easily pick up for gifts in addition to flowers.

I knew Uncle Jim would be making his wonderful concoction of egg nog for Aunt Julia so I decided to serve white wine and red punch. The colors were perfect, and the shop looked more festive than I had ever seen it before. I knew Abbey had more to do with the decorations than anyone else there. The hours advertised were from three to seven o'clock, and with the decent weather the crowd started arriving promptly. The carolers arrived in their authentic Victorian dress and were most gracious to customers going in and out of the shop. They occasionally came in to get warm, and Sarah was bringing them refreshments. The man who typically roasted chestnuts on the weekend was off, thank goodness, or the smoke would have driven some away.

"This is quite a party, Anne," said a familiar voice from behind me.

I turned around and saw Ted. It took me by surprise and all I could think of to say was, "Oh, Ted."

"I needed a couple of hospitality gifts so I thought I'd see what you had," he explained calmly. "I don't recall you having all this other Christmas merchandise before."

"You're right about that, nor have I ever had a Christmas Open House before," I answered nervously. "It seemed like a good idea when Aunt Julia decided to plan hers."

"Yeah, I was just over there and got to visit with Jim who was serving his famous egg nog," he said with a smile. "I forgot how darn good it was." Oh great, he was referring to the time when we were dating.

"Her shop is quite nice, don't you think?" I said to change the subject.

"It really is and my, how Sarah has really grown up," he noticed. "I bet she has boyfriends all over the place."

"Things change!" I said smiling as I took his arrangements and put them on the counter. "Sally, would you please help Ted? I need to get back to Mrs. Reynolds. Thanks for coming by Ted." He looked like he had more to say and wasn't going to tell me good-bye.

"Anne," he said taking a deep breath. "Can I just call you to have coffee or lunch some time? It is the holidays, and I would really like that." Oh dear heaven, how did I let him have this much time to ask me that.

"I'm so busy with this time of year, Ted," I said with a tone that said NO WAY! "Perhaps I can after the first of the year when things calm down." Why did I have to say that?

"I'll remember that. I hope you and Sammy have a very Merry Christmas!"

"Thanks, and you do the same!" I answered sincerely. When he turned to leave, I wondered how he knew I called my son Sammy. Hmmm.

Business was so brisk; we didn't close the door until eight-thirty.

"Whose idea was it to have this open mad house anyway?" Sally teased. "Do you know how much money we missed out on all these years by NOT having an open house? This was amazing! Did you also happen to see the special orders that came just from tonight?" I nodded with a big grin.

"A jolly good time was had by all, except for my tootsies, which are barking at me," Jean said as she sat down.

"Well, we didn't know what to expect, that's for sure!" I said as I gazed at the receipts.

"Like the unexpected visit of Ted Collins?" Sally teased.

"Yes, like that," I curtly answered back. "He did buy something, at least."

"You engaged in quite a lengthy conversation with him which floored me," Sally observed. "Spill the beans, girlfriend."

"I am not your girlfriend, I am your boss," I teased. "I was merely talking to a customer. What was I supposed to do?" They knew my true feelings about Ted and loved to harass me about it.

"Oh, Miss Anne, you are a dear," Jean said. "Pay no attention to Miss Busy Body over there!" We all laughed with exhaustion.

# CHAPTER 15

Mother was the last hold out to agree to go to the church tour. She didn't like the idea of leaving Harry, but I told her I had asked Mrs. Carter next door to check on him so it eased her mind a bit.

Sammy was certainly distracted when he saw the other children at the Barrister Nursery. It was going to be his first time to be left with other children. A nice lady named Bella greeted us with open arms. I stayed a bit until he became interested in a particular toy, then I made my exit. He quickly discovered I was leaving and his mouth puckered as if he knew that he should cry, but he really wanted to check out this toy. I waved the same bye-bye as I did every day so that he knew I would return. As I got in my car, a sad feeling came over me even though I knew it was a good thing.

Sue had removed her two children's car seats before she picked me up to make room for Aunt Julia, Mother, and Ella.

We were off for a day of chatter, beautiful scenery, and some spiritual Christmas joy! It didn't take long before we were off the interstate and onto curvy roads in the countryside. The winter sun and no threat of bad weather helped us relax and look forward to what was ahead. Sue had brought her brochure, which had a very easy map of the area for us to follow. I was amazed at the number of tiny towns along the route and wondered about their history. Mother said that church records in those days were essential to their town's history and ancestral heritage. She told us that Harry had connections to some of the German folks who settled in this area.

The first church was mighty small just as the brochure described. It's stately, freshly painted steeple was like one out of a story book. Inside, live decorated trees lined its walls and sanctuary. Its feature was a hand-made crèche displayed at the front altar. One of members of the congregation had made it out of paper mache. It was quite impressive with amazing details. The volunteer guide explained it had been used in the church for many generations. In the vestibule, there were homemade cinnamon rolls and fruit for us to enjoy. The smell was irresistible, and the church ladies were incredibly charming. We were ready for a little treat so we didn't want to disappoint them. They would have kept us there all afternoon if they had had their way. They couldn't thank us enough for coming. I'm sure our monetary contribution didn't disappoint them.

We left with great anticipation and went to the next church which was just three miles down the road. Along the way, we were amused by the well-maintained farms with freshly painted fences. The barns were intriguing, and the red ones stood out in the gray of the winter season. There were scenes that reminded me of the beautiful barns of Wisconsin that Sam and I had seen. It made me wonder why I didn't get to the country more often. Even in the dead of winter, the beauty shined through.

"I have always loved old barns," Aunt Julia confessed.

"I have too," chimed in Sue. "When I drive up to see Mom and Dad, I see wonderful historic barns. They're now protected and can't be torn down, thank goodness."

"Yes, we saw the same thing when Sam and I were in Door County, Wisconsin," I noted. "I don't think all states have that protection for them, however."

"Here we are!" announced Sue as she pulled in the gravel parking lot. "This is the one they say is heated with a wood stove."

"It certainly can't hold more than thirty people," Mother said as we stood in the entry area. "It's plenty warm in here for sure!"

*This church was more of a charming chapel, than a church,* I surmised.

Live greenery, rather than artificial decorations, added historic charm. It made me realize how important it was that our Main Street had chosen live greenery as well. It made a big difference in a historic building. Even the wreaths on the door were handmade with real pinecones and fruit. I paid close attention to what greenery they were using. Most were natural elements they could find in their back yard. How fun it would be to have an outdoor area where everyone could create their own wreaths from what was on hand.

Their small tree was cedar, so combined with the burning wood in the stove, there was a holiday smell, I would never forget. It made you travel back in time. I felt as if we should kneel and pray when we came close to the tiny altar. On a side church bench, lovely Christmas carols were being sung by two teenage girls with a guitar. I could have sat down and listened to them all afternoon. When we left, they handed us a small gift bag of cookies that we could munch on along the tour. These ladies knew what they were doing to give us an old-fashioned Christmas. Of course, we had to open the gift bags right away and compare the cookie assortment. This was fun!

The next church was the biggest of them all but not large by any means. It was quite old, and the tour ladies said they had recently restored it and added on a building they could use for fellowship. You could still smell the new paint. All of the church ladies were so proud of their addition and couldn't do enough to make us feel welcome.

"Have you tasted the real German coffee cake or, as we say in German, kuchen?" one lady asked wearing a fresh white apron.

"Not like this!" I answered as she handed me a cup of coffee to enjoy with the coffee cake. Four large cookie sheets of pre-sliced coffee cake were laid out before us to try.

"We have peanut butter, apricot, apple crumb, and coconut," she said. "These old recipes make quite a bit, and the dough has to rise and re-rise before the toppings can be put on to bake. The peanut butter ones are baked plain and then iced. These coffee cakes are made generation after generation."

I wanted to try them all!

"Oh, Anne, this is so delicious," Ella said with a piece of the apricot cake in her hand. "I wonder if they even have a recipe."

"The peanut butter is divine," said Sue taking big bites. "My little girl Mia would love this, too!"

"Then you shall have to wrap her up a piece and take it with you," the kind lady offered.

Sue was delighted and didn't waste any time doing just that. We all wanted to do the same but kept our manners intact. I actually did think about putting an extra piece in my purse.

We continued on and visited more churches, but the last church was the most impressive. It was small like the others, but it had a tiny curved balcony with an old organ on the second floor. The neatly carved staircase curved as you made your way up to the organ and a few pews, perhaps where a tiny choir sat. I sat for a moment in one of the pews to imagine a small wedding, baptism, or sermon that could take place here. One could feel very spiritual in prayer, right here.

To our amazement, the church's gigantic cedar Christmas tree went from the floor to the top of the ceiling, so you could see up close the historic ornaments, touch the prickly branches, and enjoy the smell of the fresh cedar. It was much bigger than my tree at 333 Lincoln. I couldn't imagine how in the world they managed to get this big tree in this little church. I wanted to sit and write a children's story book right here and call it The Teeny Tiny Church with a giant of a tree.

We spent the most time here since we had so many questions and needed to photograph every detail. A nice gentleman volunteer took our group picture as we stood on the balcony. We were offered a paper cup of hot cider when we left the church. It was a nice touch as we entered the cold outdoors and went back to our car. I left a generous contribution as if it had been the best prize of all.

The last stop on the tour was not a church, but a Museum and Heritage Center that offered a Christmas tree exhibit. Time was getting away from us, but we all agreed it was close enough to the last church that we needed to see it. It was a fairly new structure that had been added onto an old church. A sense of pride and love of history greeted us as we walked in. The volunteers were so happy to see us and immediately handed us a pamphlet that described each of the trees on display. It was like entering a winter wonderland of trees and candy for the eyes.

My attention went quickly down the row of trees where I spotted one decorated in all red and white. How could someone collect so many unique red and white ornaments? I wanted to meet that person and then steal her tree. I took several pictures to remind myself this would be a nice idea to decorate a small tree in my red and white guest room. Helen and Pat would be impressed if they came to stay there this Christmas.

Mother's favorite was a tree decorated in all bells. They looked to be quite old and must have been someone's collection. I think she examined each and every one of them. Sue had to pull me by the arm to move on as our travel time was being discussed. Each of us had our favorite tree and had hopes of remembering some of the ornament ideas.

As we headed home, we couldn't stop chattering about which church we enjoyed the most and then which treat was our most favorite. I told Ella that we needed to figure out this wonderful German coffee cake recipe. We all agreed it was a spiritual experience as well. Our Heavenly Father must be quite impressed with all the different ways we celebrate his birth. I would never be looking at churches the same way again. It was God's art created by man.

# CHAPTER 16

As we got closer to home, Ella and I wondered how our Sammy was doing on his first play day. We dropped off Mother first since she was the most anxious to get back.

"I'm so glad you got her out today, Anne," said Aunt Julia after Mother got out of the car. "This is what happens when one of the two has health issues. It takes down both of them."

"I can see it happening in my own parents," added Sue. "I feel badly that I'm not there to help them at times."

"Yes, it's what I admire so much about Sam's sister, Pat," I recalled. "She has always been there for Helen, and Sam used to feel so guilty about that. She never complains about it either!"

"That's wonderful, Anne," commented Ella.

After we dropped off Aunt Julia, Ella and I thanked Sue for her most unique Christmas gift.

"You are most welcome," she responded. "I'm sorry I didn't get to your open house at the shop, Anne, but I was so wiped out that day and so were my little guys."

"Of course, Sue, I understand," I said getting out of the car. "We'll be seeing you more through the holidays, I'm sure!"

Ella went in to the house, and I left to rescue Sammy from the Barrister Nursery. When I arrived, he was sitting next to a cute, chubby, little girl that looked to be around the same age. He was quite taken with her, and it was a precious Kodak moment I wished I had a picture of. When he turned to see me, happy fireworks exploded on his face. We hugged and kissed as if we hadn't seen each other for a long time. After hearing a nice report from Bella, I decided that Sammy would have play days more often.

That evening, after Sammy was put to bed, Ella and I sat at the kitchen table with a glass of wine and discussed the list of invitees for the Christmas party. I wanted the event to be spectacular and suggested we book Chef Michael, a chef that Sam had hired for our private New Year's Eve dinner last year. Ella reminded me it would cost a fortune, but I told her I wanted to serve food that would be different than every other Colebridge party. I had purchased some lovely invitations from The Written Word, and if Sarah had time, I would have her hand address them with her beautiful hand writing.

"I should have sent these a week ago," I told Ella.

"Trust me, Anne, they all know about the Dickson Christmas party and have it on their calendar, whenever it might be," Ella teased with pleasure. "No one will miss this party! Still no word from Mrs. Dickson?"

"No, but everything will be ready for them, even if they decide to come at the last minute," I assured Ella. "I sure think a little red and white Christmas tree in that guest room would be a nice touch if they come."

"I suppose you could do it anyway," Ella suggested. "You said you wanted it all to be ready."

"Ella, I need to ask your advice about something," I said looking serious. "Nancy and Richard asked me to go to a fundraiser that's a wine tasting gala. I feel I should go because they are chairing the event." I paused before continuing. "Would you be available to stay with Sammy if I go? I would put him to bed first, although I think it starts before then. I just hate to tell them no, and I could easily leave after it gets going."

"You absolutely should go," Ella said firmly. "I have no plans other than my usual TV shows so you should have an evening out with those nice folks." I felt at ease with her response and decided I would try to make it.

The next morning was Saturday, and my plans were to stop by the shop to check on things and then take my walk along the trail. It would take me by the train depot where the Christmas Quilt show was being held. I was anxious to see the exhibit and especially my quilt.

I went into my calm, quiet shop and, as always, the flowers were a breath of fresh air, especially in the winter time. I went into my office and quickly scanned my emails. Ted's email, as always, popped up from the others.

*Good to see you happy and busy last week Anne. Look forward to having that cup of coffee or lunch you promised. Season's best, Ted.*

I knew that wasn't going to happen, but I couldn't be rude to him at my open house where he was buying something. I didn't respond. It gave me almost a creepy feeling that he was still interested in me.

It was a very cold and cloudy morning as I briskly walked along. They say you burn more calories in the colder weather so that was my only consolation as I endured the cold wind against my face. In a bit, the depot and the Mistletoe Market would be open in the park. When I arrived at the depot, the shelter from the wind near the river felt good, despite the place not being heated. Sharon, who worked for Phil, was moderating the quilt show when I walked in.

"Wow, this is spectacular," I said seeing the walls of red and green that were practically yelling Merry Christmas.

"This idea worked out nicely," Sharon said looking at the quilts. "It's sure a lot easier than hanging all those quilts on buildings, I can tell you that!" I laughed in agreement. "There's your Lone Star quilt." She pointed to the back wall of the depot, where it shined brightly. "I wish you could hear all the great comments from people."

"It's stunning there," I said coming closer to see it. "I'll have to send my mother-in-law a photo if she doesn't come to visit this Christmas. I'm surprised there hasn't been a quilt show here before!" I got out my phone and decided how closely I should take a photo.

"There aren't many events here at all because it isn't air-conditioned or heated," Sharon explained. "At least people have their coats on when they come in and seem pretty contented about it. They also love that it's a Christmas quilt show. I've never been to one, have you?" I shook my head.

I stayed longer than I realized. I couldn't believe they had gathered so many Christmas quilts. There was one row that had only miniature red and green quilts. The workmanship amazed me. It was awesome to see them all in one place. The accents of live, green roping and trees made it quite the quilt wonderland.

Abbey was working with prospective customers when I walked in the Market tent. The tent was warmer than I expected it to be, but the bright sun was now warming up the temperature.

"I see you have all the trees lit today," I noted.

"Well, they all weren't lit this morning when I got here," she admitted with frustration. "Kevin spent most of the night here, but he got too cold and decided to go home around three o'clock in the morning. I wonder if someone was watching until after he left. We couldn't afford a reliable camera to video all night, plus we just have a couple of more weekends to go."

"You should have stayed here with him to keep him warm," I teased.

"I did, after he ended up at my place in the middle of the night," she said with a wink.

"I asked for that, I guess," I said winking back.

"Well, don't just stand there, try to find something to buy," she said, giving me a nudge on the arm.

So, I did just that. The first thing I went for were the beautifully decorated Christmas cookies before they were all gone. I knew later in the day, they would all be sold. There were at least seven different designs decorated in such detail. I couldn't believe it. *These should not be eaten*, I thought.

When I took them all, Abbey gave me a dirty look. I

headed to the rack of handmade crocheted scarfs. There were many bright colors to choose from. I picked out several, knowing one would go to Ella for sure. Displayed under one of the trees was a group of teddy bears that were made out of old quilts. I spotted one in red and white that was adorable and put it on my pile to purchase.

"Isabella had a fit when she saw the bears made out of old quilts," Abbey reported. "She said no one cuts up a quilt under any circumstances."

"Well, he's going to have to go home with someone," I stated, as I held him close. "He will be well loved at my house, so home it goes! You've done a great job here, Abbey. I hope your idea of owning a Christmas shop will be a reality some day!"

"Well, I better win the lottery then!" she quipped.

# CHAPTER 17

As I was getting dressed for the wine gala, I felt strange and even a bit guilty in dressing up for what was not a family affair. This was a social event in Colebridge, and I would be attending as the widow of Sam Dickson.

I did as Nancy suggested and put on a simple, black dress that had a deep neckline to show off my pearl necklace and matching accessories. I couldn't go wrong with this classic look. *I must have every kind of black dress known to mankind*, I thought. I hadn't worn high heels for some time, and it felt awkward. I pulled my hair back for a sleeker look. I looked in the mirror and could almost hear Sam whistling at me.

I said out loud, "Honey, I hope you would approve of me going out tonight. I miss you!"

I put Sammy in his crib for the night, and the look he was giving me told me he knew something was different. I'm sure he was wondering why his mommy was wearing funny clothes. I turned on his little music box and left his room feeling a bit strange once again. Did I want his approval to have a night on the town as some would describe it?

I put on the full-length, fake fur coat Sam had given me before one of the Martingale Christmas parties. Ella was watching and told me that I looked like a million dollars and to have a great time.

I took advantage of the valet parking and walked into a beautifully lit room of total elegance. I'm sure Nancy saw to every detail to get the atmosphere she wanted. A handsome young man in a tuxedo took my coat and another attractive lady said she would escort me to my table that had the Barrister name on it.

I saw the table had cards for the place settings as if we were going to have a formal, gourmet meal.

"There you are gorgeous," Richard greeted me with a big smile. "I was beginning to wonder if you were coming! They'll be starting any minute. When everyone is seated, I'll introduce you to the rest of the guests at the table."

"Oh, that's great!" I said seeing my name on the place card.

"Anne, you're here!" Nancy said approaching me in a Christmas red gown that made her look like Marilyn Monroe with her blonde hair and revealing cleavage. She gave me one of her endearing hugs.

"I hope you get a chance later to see the wonderful silent auction items," she noted. "Thanks again for the generous gift certificate."

"You're welcome," I responded. "I'm glad I could support this worthy cause. The decorations are so beautiful Nancy. It must have cost a fortune."

"That's why you need to spend a lot of money tonight, Anne," Nancy teased.

At last, the emcee announced for everyone to take their chair and everyone was seated. Richard then arose from his seat and went around the table to introduce everyone. "This is Ginny and Irv Whitman, next to them, looking so beautiful, is Marlene Hopfer and Melba Bachmann. Then we have the distinguished Jack Fletcher, and of course, you know Nancy and me. Next to me, it's my good fortune to have Anne Dickson."

We all said hello to one another before Richard explained how the evening would flow. Every woman at our table, except Nancy, chose to wear their black dress just as I had.

"With each course, there will be a tasting of a special wine that the emcee will describe to you," Richard began. "As you proceed with the evening, you can make small notations of your favorite wine on that small pad we have put by your place setting. They'll be happy to sell you a bottle or a case of your favorite wine at the end of the evening with all the sales benefiting our charity here this evening."

The emcee, Dan Borger, began with a five-minute description of the worthy organization and thanked the committee and Mr. and Mrs. Richard Barrister for acting as chairs of the event. Much applause followed as Richard and Nancy stood and responded to the recognition. We began with light conversation among ourselves as the first appetizer was served.

"You seem to have captured the attention of that handsome man across the table from you," Nancy whispered as she leaned forward. Haven't you noticed?" I was too embarrassed to look but assumed she meant Mr. Fletcher.

"I have to admit he is candy for the eyes, as my friend Abbey would say," I quietly whispered to her. "Don't go where I think you are going Mrs. Barrister, and mind your own business."

"I'm just saying that he's single, and I'll not say any more!" Nancy said as she took a swallow of her wine.

I ignored her, and now, especially Mr. Fletcher because I didn't want to appear to be interested. Instead I now engaged in conversation with the two attractive ladies, Marlene and Melba, who worked for the Barrister Funeral Home. Exchanges went around each table as we enjoyed the courses. They were light and not filling, but trying all the wines was a new experience for me. The red wines were definitely my favorite and I found the information to be very interesting. I had always wished I knew a little more about wine.

When the last course was served, they announced that it was called Chocolate Paradise—and that it was. The Chocolate mousse was placed on a round chocolate cake, followed by chocolate whipped cream, adorned with chocolate shavings and a few raspberries. It got even better when the complimentary wine was a merlot called Red Devil. The emcee joked that it was both heaven and hell brought together to end the evening. He was right about that. Chocolate and merlot together was always a favorite of mine.

The merlot was heavy and mellow, just as I liked it. Mr. Fletcher had the same positive reaction, but some of the others felt that it was all too sweet. Feeling much more relaxed and mellow, I made a toast to our generous hosts for the evening and then to my favorite wine of the evening, which seemed to tickle Nancy as she saw me enjoy my favorite combination.

Mr. Fletcher's eyes met mine for a single moment as he was the first to add "here, here" to my toast. I made a note of the Red Devil on my note pad so I could purchase a case of it before I went home. A bottle of this merlot would be a good Christmas gift for many, and I would be sure to serve it at our Christmas party at 333 Lincoln.

"Nancy, all this wine is a bit much," I said feeling light headed. "Even though it's been small sips of each, it's a lot of wine which I'm not used to." She laughed and said I should walk it off by taking a look at the auction items.

I did just that after I went to the wine station to put my name in for a case of Red Devil. As I gazed at the many things that I did not need, nor even want, I did see my gift certificate. I happen to notice that the bid had gone way past the hundred dollar value which I thought to be strange.

"I hope you don't bid on that Anne because I'm determined to win that one," said Jack Fletcher pressing next to me in the crowed line. He caught me by surprise.

"Did you notice that you are way above the original value?" I said, now feeling stupid at the remark. "You know it's a shop here in town where you could buy directly!" He laughed.

"Well, it's because I want them to make money this evening for a worthy cause, and I won't be doing that if I come and spend it in your shop," he said looking deep into my eyes.

Of course, he knew it was my shop. I felt stupid once again as if he had to explain his actions to a child. I could feel my face turn red.

"Oh, I wasn't sure you knew this was my business," I said shyly.

"Of course, I knew it was your shop," he said with a flirtatious tone. "You have a wonderful flower shop and excellent taste in wine, I might add."

"Oh, so you are ordering the Red Devil, as well?" I asked awkwardly.

"I am, shall we pick up our orders together?" he asked. "I'm getting ready to leave, and if you need any assistance, I will be glad to help you with it."

"Oh, no, I'll be fine," I said gathering my senses. "I have placed my order, and they will carry it to the car when I'm ready to leave. I'll leave you to finish your bidding. You may want to add a bit more to that." He laughed as I walked away toward the other auction items.

After saying hello to some of the folks I recognized, I suddenly realized that no one was driving me home and that I had to compose myself. Fortunately, the case of the Red Devil and I made it home safely, but near the midnight hour which truly surprised me. I had no intention of staying that late.

I peeked in on Sammy, and he was lying on top of his cover just like his Father used to do. I took off my dress before remembering to take off all my pearls. I was relaxed and my mind was not quite as in order, as I was used to. I wanted to chuckle to myself. I didn't seem quite in control. Was it the wine?

When I slipped between the sheets, I played back the encounter with Jack Fletcher. I hated to admit that he was, indeed, attractive, and I especially liked his dry sense of humor. He did seem to like Brown's Botanical Flower Shop so he couldn't be all bad. For once in a long, long time, I fell asleep with a smile on my face.

# CHAPTER 18

When I came downstairs the next morning, Ella had already rescued Sammy from his nursery and brought him down to breakfast. I slept later than usual, and there was no doubt that it was from the wine I had enjoyed the night before.

"I see someone went shopping for some wine last night," Ella said with a smile in her voice as she looked at the case of Red Devil.

"It was for a good cause, my dear Ella, and it happens to have a wonderful taste especially with chocolate," I explained humorously.

"You had a good time then?" Ella asked as she poured me a cup of coffee.

"I really did which surprised me," I began to explain. "It was such a unique fund-raiser, and half the town of Colebridge was there, for sure! I saw a lot of my customers so

I'm glad I contributed. I'll have you know the hundred dollar gift certificate I donated from the shop had over bids on it."

"Well, my goodness sake," responded Ella as she gave Sammy more Cheerios on his tray. "Our little fella here slept all night without a peep. What are your plans today? Should I prepare dinner?"

"It's pretty cloudy out, but I thought I would take Sammy with me this morning to Sam's grave site and then stop by Sue's house for a visit," I described. "I think I need to get Sammy around other children more often. Do you have plans?" I wondered if Ella ever attended church, but I felt awkward in asking.

"Well, if you'll have Sammy with you, then I'll stop by the hospital to see a friend of mine," she said wiping Sammy's mouth. "She doesn't have much family to come around."

"You absolutely should, Ella," I stated. "I have the stroller in the car, and thought if the weather cleared up, we may take a stroll on Main Street as well."

"You better bundle him up good," Ella warned. "We've been lucky that he hasn't caught as much as even a sniffle this year."

"I'll be sure to do that," I smiled. "He's getting good care by a special person so he's been fortunate." Ella smiled looking at Sammy.

I was always taught that Sunday was a special day to be thankful and think of those you love. It was usually a day Sam and I would have spent together away from our busy schedules during the week. I felt the urge to acknowledge him in some way so off to the cemetery we went.

I opened the stroller on the graveled path at the cemetery and secured Sammy inside with a blanket on top of him. So far, he seemed to think it all quite fun. We stopped at Sam's grave site, and Sammy looked back at me as if he were to say, "Now what? Can't we keep going?"

"Sammy, can you say 'Da Da?' I asked, thinking I may have some luck. I had tried to relate Da Da to a photo of Sam at home for Sammy to relate to. Sammy was looking somewhat confused. "Your Daddy has been laid to rest here, and now he is up in heaven looking down on you and me," I explained as if he would understand. "Your Da Da liked trains, just like you, your Grandmother said. You know, like in the choo choo book." He then attempted to say choo, as he slobbered from his mouth. "That's right choo, choo."

He then started squirming and kicking forward as if to tell me to keep the stroller moving.

"Can you throw Daddy a kiss like you do Mommy?" I asked with hope. He shook his head, like he did when he didn't want to do something. He was anxious to move on.

"Okay, we'll go then. I just wanted you to say Da Da for your father to hear."

I turned the stroller around to head back to the car, and what did I now surprisingly hear? The words "Da Da" were said loud and clear. I stopped to give him a kiss on the cheek. I couldn't believe it. What a cool happening!

"Did you hear that, Sam?" I said aloud. "What a sweet boy you are, Sammy!" I hugged him dearly before putting him back in the car seat.

When we arrived at Sue's, I couldn't wait to tell her about our visit to Sam's grave site.

"You are a crazy Ma Ma, you know that?" Sue teased. "Would you like a hot cup of tea?"

"Sounds great," I said taking off Sammy's coat.

Sue told me Eli was asleep, but Mia made her presence known immediately and couldn't wait to check out the "baby," which is what she called Sammy.

"I'm glad you were happy with Barrister Nursery," Sue said as she tried to get Mia to settle down.

"Oh, Bella is a dear," I bragged. "I could tell the little ones like her as well."

"I don't know if I could work anywhere else after having this convenience in the same building," Sue confessed.

"So work is going well?" I asked to catch up.

"I should respond by saying yes, but it is an occupation of grief," she reminded. "Sometimes it's hard to forget people's sadness at the end of the day. The holidays are coming which makes everything worse."

"I can relate to that," I said taking a sip of my tea. "I can't wait to get through it!"

"I'm glad you stopped by because I have a situation I am struggling with that I will share with you in confidence," Sue said looking down at the floor as I waited to hear. "I have this one gentleman in my group who lost his wife suddenly a few months ago, and he's very regular about attending our meetings. I enjoy his company very much because he's intelligent, successful, and very kind. He's average looking, I suppose, and a bit older, I'd say. There's no question that he has taken a special interest in me which I try not to let happen, but I'm afraid the feeling is mutual." I was surprised this was happening to Sue. I didn't respond because I wanted to hear more.

"His wife never wanted children, and because he loved her dearly, he accepted the idea. He talks quite highly of her, and I know he misses her. He really likes to talk about my life with Eli and Mia. I think it's his dream of what he would have liked with his wife. He met Eli and Mia one day when I was heading to the nursery. He hangs onto every word I say, which is a new experience for me. Long story short, Anne, he wants to take me to dinner. It's not against the rules, but it really isn't a good idea professionally." She took a deep sigh. It was my turn to react.

"I understand," I said, trying to absorb it all.

"There's no question that he's probably on the rebound," Sue also noted.

"Makes sense," I finally said when she looked at me.

"The problem is that I really do want to go to dinner with him, and after all it is the Christmas season," she admitted. "We both want to do something special, I suppose." She took another deep sigh.

"So Sue, how have you responded to him?" I felt sad for her.

"Well, I tried to explain to him what he's going through, and why I shouldn't accept," she said with a frustrating tone.

"He's not buying it, right?" I had surmised.

"No, he's not, and right now, I'm not sure I am either," she said laughing.

# CHAPTER 19

Sammy was entertained with new toys as we continued to sip our tea. Eli woke up, and when I picked him up to hold him, Sammy's face puckered in disappointment. As Sammy loudly expressed his opposition, Eli also began to cry. We couldn't help but laugh.

When things settled down a bit, Sue and I revisited her situation.

"It's just too soon for him to be thinking about any kind of a date, with me or anyone else," Sue stated.

"But it's not too soon for him to have an understanding friend," I said smiling in approval.

"True, but there's another factor to consider, and I don't want you to take this in the wrong way," she said in a serious tone. What was she getting at?

"Since his wife died, and he is a widower, the spouse who is left behind has a tendency to put the deceased spouse on a pedestal," she carefully explained. "That person could have even been unfaithful or a difficult person to be married to, but none the less, they have a tendency to think their spouse walked on water. It's hard to fill those kinds of shoes."

I thought about Sam, of course. No one could fill his shoes, but he was one of those exceptionally good men. I knew she had a valid point.

"So what does one do to avoid that?" I asked not sure I wanted to know the answer.

"I tell them that no one will ever replace their spouse, and that no one should make any comparisons," she advised. "That's very hard to do, especially when you've only been married to one person. There's a certain amount of guilt that sometimes goes with giving the deceased this pedestal. It's very hard to move on if you want to play a martyr. One can miss out on a whole new life." I stared at her. "Oh, Anne, I wasn't referring to you in any way, but I do think you need to be aware of what can happen. I see this more in women than men." I nodded feeling rather sad and bewildered.

"The guilt in moving on is not good either, right?" I asked thinking of my reaction to Jack Fletcher. "I felt guilty last night just having a good time at the Wine Gala I went to."

"That's perfectly normal, Anne," Sue revealed. "Don't you think Sam would want you to enjoy life and be with people you enjoy? He had a hundred percent of you in your years together. You can't give more to anyone, and you should feel good about that."

"So, Miss Sue Davis, what have you decided to do about this dinner date?" I inquired in a teasing manner.

"I'm going, he just doesn't know it yet," she said, laughing at herself. "I'll keep it a light friendship, and we'll see what happens. I could see myself really falling for this guy so I'm the one that will have to really watch it."

"You're so wise, my dear cousin," I said with admiration. "How much do I owe you for this counseling session?" She laughed.

"How about babysitting for me when I accept this date?" she kidded.

"It's a date, my friend," I assured her. "I would enjoy that very much!"

The time had passed quickly, so I picked Sammy up off the floor and off we went. It seemed like this visit was meant to be. We helped each other play out what was on our mind as our little cousin relationship connected. I had much to ponder with her words of wisdom. Did I put Sam Dickson on a pedestal? Yup, I think he was up there for sure, and I couldn't imagine when he wouldn't be.

# CHAPTER 20

The Jane Austen Christmas Tea had Jean preoccupied most of the time even when she was at the shop. Unfortunately, with such a small club, the majority of planning fell on just a few. My responsibility was centerpieces for the tables, and I was contributing a Christmas Tea Pot for a door prize. It was corny and yet adorable. Nancy volunteered to work with Donna on the food so I knew it would be delicious, attractive, and even true to the time line of Jane Austen. Donna assured us her place would be decorated with live greenery instead of her usual, artificial décor for the Christmas season.

Donna was especially excited about showing her antique feather tree each year. It was the world's first artificial Christmas tree, which originally came from Germany. Hers was a four foot, 19th century tree that would be decorated with her collection of antique ornaments. The name feather could be deceiving in that they really were green dyed feathers wrapped around wire to resemble pine needles on the branch. They became very fragile over time. The tree was constructed to be able to bend the branches upward when not in use, but as they aged, the branches had to be left down, or they could break off. This no doubt would be a conversation piece as guests entered the tea room. Donna said that she would give a brief history on the tree at the tea which I thought was a good idea.

Donna admitted that it would take two or three sets of her Havilland china to accommodate our attendance at the tea. She also told us that it would require hand-washing and drying for which the club had agreed to pay her servers. We were charging a hefty admission so we could at least break even with our expenses. Jean, as the hostess, would be wearing her Jane Austen attire, and Donna's servers would be wearing white pinafores with matching dust caps. Everything seemed to be in place except the unpredictable Missouri weather. They were calling for snow the afternoon of the tea. Mother had already called to see if there was word of the tea being cancelled.

"It would take a tornado to ruin this day for Jean," I warned her. "You may as well plan on getting dressed and drag out that fur coat and boots of yours. It will be fun!" She was turning into such a worry wart these days.

"When do you think it will be over?" Mother asked in her voice of concern. I knew Harry was the reason for that question.

"I'd tell Harry these things usually go at least two hours, and I'll have you back in proper time," I assured her.

"Okay then," she agreed. "I'll be ready."

The shop was going to be shorthanded today. Jean was off for the tea, and Abbey could only fill in for her a few hours since she had to get back to the Mistletoe Market. I had planned to stay at the shop till it was time to pick up Mother. It was at times like this that I wished I had another employee. Before Sue had her children, she was my back-up employee. I knew Aunt Julia would be attending the tea and would be leaving Sarah at the shop by herself.

When I arrived at the shop, Sally told me that a man had called and asked if I would be in the shop today.

"Well, if he didn't leave a name, it's probably a salesman who just wants to drop in on me unannounced," I surmised. "I hate that."

"I told him you were scheduled to work tomorrow so I hope that was okay," Sally noted as she put out more Christmas ornaments. I nodded.

When I saw the order sheet, I knew I had better get right to work instead of checking my emails. Orders were coming in for Mr. Heflin, a retired school teacher in Colebridge, who had passed away. It was strange how owning a florist put you in the line of the earliest notifications when someone died. Deaths around the holidays were extra sensitive. I always got more specific about what the customer wanted for the arrangement so we didn't create something too festive.

Aunt Julia came in the door with a frown on her face and with a grumpy good morning greeting. She looked around to see if anyone was in the shop before she started complaining about shoplifting in her store.

"Do you all have that problem here?" she asked curiously.

"Not really," answering as I thought. "Sally mentions now and then that she suspects greeting cards and an occasional ornament gets taken, but nothing we could ever prove. You know this is the worst time of year for shoplifting, Aunt Julia, especially with a shop like yours. You have some nifty small things folks can put right into their pocket. All those goodies of yours would make great Christmas gifts."

"Well, do I call the police or what?" she asked in confusion.

"It's up to you, but I would call George for sure so he can put an e-blast out to the merchants that this is occurring," I suggested. "It reminds shop-keepers to keep an eye out."

"Good idea," she said with a nod. "He sent one out a couple of weeks ago about the mystery wire cuter. Did they ever find out who that was?"

"Abbey hasn't said, and I'm sure she'd tell us," I answered as I kept working.

"Well, I'm going to run on home to change for the tea," Aunt Julia said going towards the door.

"If Sarah's alone there today, you might want to remind her to stay off the cell phone and be extra watchful," I said without thinking. "Sorry, Aunt Julia, I don't mean to interfere."

"No worries, we've already had that conversation," she said waving good-bye.

# CHAPTER 21

After freshening up, I felt I looked appropriate in my tea attire. I decided to wear my winter white skirt and sweater so I could wear my usual pearls.

I thought of the MERRY CHRISTMAS JANE poem that I had written at one of our planning meetings for the tea. It was corny but to the point. It went like this:

Oh, I wish that I could see, Jane Austen
at this Christmas Tea.
I'd hope for pudding, pie, and cake,
which she would serve and claim to make.
A gift of paper and pen I'd give,
so her prose of words would live.
Merry Christmas, Jane, I'd say.
I hope you enjoyed our tea today!

When Mother and I arrived at Donna's Tea Room, we were greeted by Jean, like every other guest. She looked very

English and like a relative of Jane Austen's. She took my door prize to put with the others before I hung up my coat. Donna's feather tree was getting a lot of attention, and she was happily answering everyone's questions as they admired it.

Fresh cedar garlands accented with pink velvet bows were the perfect choice to compliment Donna's personal taste. Under each chandelier was holly and mistletoe with a pink bow attached as well. The flower shop's holly centerpieces were small and delicate to make as much room as possible on the table. Donna's Havilland china patterns, with touches of pink, were well chosen and had been placed ahead of time. The antique white napkins from her linen collection were each different and had a small piece of holly and pink bow around them. *It's all in the details,* I reminded myself, *and women love it.* I couldn't help but wonder who had to wash and iron all these lovely linens. Donna always said that no matter how expensive her collections were, they were made to be used. Sharing her things with others gave her great joy.

As soon as Mother and I took our seats, we were joined by Sue and a sweet, elderly lady wearing a hat and white gloves. One of the servers immediately poured our tea and offered sugar cubes with tiny pink roses on them. I wanted to take one just for a souvenir they were so precious.

Jean very quickly got our attention. "Jolly good afternoon, ladies," she began. "Welcome to our Jane Austen Christmas Tea Party. A friendship tea, such as this, is a proper note to prepare your heart and home for the holidays ahead. Let us raise our cups and toast good health and happy spirits to all!"

Everyone happily did just that as they responded with wishes to each other around the room.

"While you enjoy the special delights Donna has prepared for you, I'll share with you a few words about tea," she continued. "We have Anna, the Duchess of Bedford, to thank for the proper tradition of afternoon tea. In her day, the English typically ate a large breakfast and had a late dinner, without much nourishment in between. When the Duchess experienced a sinking feeling during the afternoon, she would treat herself to tea, cakes, and biscuits around four o'clock. Thus was born the custom of taking afternoon tea, as you are today, at four." Everyone clapped and chattered a bit. "For a bit of humor today, I thought I would share with you a few bits of superstitions regarding tea." The room silenced.

"If you find bubbles on the surface of your teacup, it is said that you'll receive money," she said with a giggle in her voice. "If loose tea leaves float to the top of your cup, it is rumored that a stranger will come to visit." She paused as their chatter became louder. "Now, if you found yourself stirring your tea in the teapot, it would mean that you are stirring up trouble." Everyone laughed. Only an English accent like Jean's could carry this off so well. "More outrageous, ladies, is if two women pour from the same pot of tea, one will give birth to a child with red hair within a year!" All eyes were on the servers as they laughed. "Let us toast our beloved Jane Austen who had a great fondness of tea." Everyone raised their tea cup and said "here, here."

Jean continued. "Jane was privileged when it came to having to do household duties, but she did write it was her responsibility to keep the tea well supplied. Christmas in the Regency era would be described as elaborately simple. It would have been fun to have been a little mouse and absorb the happenings of the holidays. Jane's past is our present to you today. Jane herself said, I quote: 'Think only of the past, as its remembrance gives you pleasure.'" Everyone clapped. "As I leave you to partake of more treats and conversation, let me acknowledge our Jane Austen Literary members, who wanted to share this tea with you today. Will you all please rise?" Everyone gave their applause as we gave a slight curtsey for the occasion.

Donna then stepped forward and acknowledged Jean's diligence and what a delight she was to work with. Of course, her comments drew more applause.

It was difficult to not have another taste of lemon curd scones, cream cheese and chutney sandwiches, lady fingers, smoked salmon and watercress on crackers, strawberry bread, and garlic cheddar biscuits. I told myself one had to be mindful of the proper occasion and not indulge on the delicacies.

While Mother was engaged in saying her good-byes, Sue leaned over to tell me her dinner date was the coming weekend. I grinned and reminded her of my willingness to watch the children. Surprisingly, I was the one that suggested it was time to leave the tea. I knew Mother had a good time as we greeted "Merry Christmas" to all her friends. I gave Jean a big hug and congratulated her on a job well done. She was special, for sure. I could tell however, she was very relieved to be seeing the guests out the door.

The Jane Austen Literary Tea was now over. A good time was had by all as we once again celebrated our dear Jane. As I let Mother out of the car, sleet started to ping on my windshield. I smiled at the good timing.

# CHAPTER 22

Christmas orders were increasing at the shop, and my attention and labor were both needed badly today. The sleet had turned to a light snow, but the temperatures would not let it stick for long. I skipped any kind of walk, with my busy day ahead, and after giving a hug to Sammy, I was off for the day.

Fortunately, the Mistletoe Market was closed today so Abbey was available to help us. She started on the Optimist Christmas Party centerpieces, and the rest of us tackled the large order from the Colebridge Country Club. In walked Nick from across the street headed to the counter where he saw Abbey.

"Hey, I have a bone to pick with the Christmas committee, Abbey," he said boldly. "Who do I talk to?"

*Oh, dear, now what,* I thought. She didn't have a chance to respond before he started to describe his complaint.

"I don't know who on earth hired the Sugar Plum Fairy character of yours, but they're asking for trouble, if you ask me," Nick said before he was cut off by Abbey,

"Okay, Nick, the committee has already received complaints about her, so you're not the first," Abbey quickly responded.

"What's she doing?" I asked jumping into the conversation. Thank goodness we had no customers in the shop.

"Hey, she can wear as little clothing as she wants, but I don't like the way she talks and treats people," Nick said in anger. "Mrs. Smithton was purchasing a nice candy order from me inside my shop while her husband waited patiently outside. Her husband comes inside and tells us that the Sugar Plum Fairy kept punching him in the belly with her wand and said, 'No more sugar for you, Daddy.' He didn't think it was very funny because he's a very sizable man, and he felt insulted. I apologized, of course."

"That sounds like her," Abbey said, shaking her head in disgust.

"How mean!" chimed in Sally.

"It's gotten worse as her comments have been sexually insinuating," Abbey added.

"Well, I say get rid of her now before she offends anyone else, don't you think?" Nick suggested.

"I'll report this to the chairman for you, Nick," Abbey offered. "I know they're dealing with the issue, but you know how it is if you fire someone. She's the type that would sue you and even make a big stink in the press with it. We also have one of the Santas that's been doing a lot of naughty pinching, too."

"What?" I asked in shock. "Is he pinching little children?"

"Oh dear, I shouldn't have said anything," Abbey replied now feeling embarrassed. "The trouble with some of these folks, who get into character, is that they think they can get by with anything to get people to laugh at them. The committee interviewed them, asked them how they would act out their character, and drug tested on them because you never know what might happen. I am new at all this, but I am told that every year there is some kind of mischief the committee has to address. The actors get paid fairly good money, but then they have to spend a lot of time out in the cold. They are told to take breaks, but not if they have big crowds around them."

"That makes my hackles rise!" said Jean with wide eyes. "Perhaps she's a crumpet, by golly!"

"Whatever that is!" laughed Abbey.

"They are looking for action to profit, shall I say," best described Jean. We all shook our head and laughed, knowing it could be true.

"Okay then, "I'll count on you taking care of it," Nick finally said, calming down.

"We really don't want this kind of information to get out, Nick, so please keep it to yourself," Abbey pleaded. "We don't need this kind of publicity."

"Well, it will stay in these botanical walls," I assured her.

"Saint Nick always frightened me as a child," shared Jean. "Do you think most children are frightened when their parents prop them on his lappy?"

"Oh, I know," agreed Sally. "The poor kid screams and mommy and daddy take a picture and think it's so cute. Oh dear, Anne, were you thinking about taking Sammy?" I laughed.

"He doesn't know it's Christmas," I said with a smile. "Perhaps this is not the year to attempt that."

"Right so, Miss Anne," Jean cheered.

# CHAPTER 23

I was nibbling on peanut butter and crackers at my desk when Sally pecked on my door to tell me there was someone to see me. I wanted to ask who, but she quickly disappeared.

I walked out of my office licking my sticky fingers and saw Jack Fletcher standing at the counter.

"Oh, hi there!" I said surprised at his presence.

"How are you?" he asked with a big smile showing his incredibly white teeth, which I remembered from when we met.

"Oh fine," I answered, not knowing what to say next.

"Have you had a chance to enjoy anymore of that Red Devil?" he said with a tease.

"Not yet," I said, feeling my face blush.

"I brought in my gift certificate today," he said holding it up.

"Great, what did you have in mind?" I asked in my professional voice.

"I spotted a couple of matching Christmas arrangements in your case here," he described as he walked over to show me.

"Sure, I'll take them out of the case for you to see better," I said as he watched me proceed.

"They're very beautiful, did you arrange these?" he asked, giving me a direct grin.

"No, I didn't. I think Abbey did these. Am I right?"

Abbey was looking at us with her mouth open. "I did, and I'm so glad you like them," she said looking him over from head to toe.

I walked behind the counter to write up his sale when he told me they would have to be delivered. When I asked where, he gave me my church's address. I wanted to ask a million more questions, but it really wasn't any of my business.

"You can tell your delivery guy to leave them with Karen in the office," he instructed.

"That's very thoughtful," I said not knowing what else to say.

"Have you had lunch, Anne?" he asked as if we were old friends.

"Oh, yes, the peanut butter was very fresh and satisfying," I said in a silly manner.

"Hard to top that," he teased back. "Perhaps we can have lunch at another time then?" I nodded.

"We'll get these delivered first thing in the morning," I said getting back to the purpose. "Our delivery guy already went home for the day."

"That'll be great," he said, as he went to the door. "Have a nice day, ladies."

As soon as he closed the door, Jean, Sally, and Abbey walked up to me at the counter, grinning at me for a response."

"What?" I said grinning back.

"You tell us!" Sally said first. "What a handsome dude, Anne. How do you know him?"

"Settle down, you guys," I said putting aside his order for delivery. "I met him at the Wine Gala because he was sitting at my table."

"Is he a single gent?" Jean asked wanting to know more.

"Nancy said he's single so now let's leave it at that, you all," I responded as I got flustered.

"Sending flowers to a church out of the goodness of his heart is pretty endearing, Anne," Sally added.

"You all chew on it," I instructed with humor. "I am going back to my peanut butter and computer."

It was hard to concentrate the rest of the afternoon. His visit threw me off in many ways. When I saw we were caught up, I decided to go home early. When I got in my car, my cell phone went off. It was Mother with a sobbing sound in her voice.

"What is it, Mother, is it Harry?" I quickly asked in a panic. "What is it?"

"Mrs. Carter passed away today from a heart attack," she finally said. "I had no idea she had any heart problems."

"Oh, I'm sorry to hear that," I responded. "Mr. Carter was the one we were always worried about."

"Yes, and this has just devastated him," Mother said sniffing. "He's been in terrible shape. Harry and I are trying to be helpful until their two sons get here from out of town."

"Do you want me to come by?" I asked feeling so sorry for her. Mrs. Carter had not been a close friend of Mother's, but she had lived there as a neighbor as long as Mother had. We counted on Mr. Carter for many things when it was just Mother and I. He hadn't been able to do much the last couple of years because of his health. It was just comforting for Mother and me to know the Carters were always there.

"No, honey, that won't be necessary," she said sounding better. "I just wanted you to know right away. I will let you know when the arrangements have been made for her. You will probably know before me anyway."

"I will be sure to send flowers from us, so don't worry about that," I assured her.

"That would be nice, Anne, just let me know what we owe," she noted.

"You are going to miss her, aren't you?" I asked before hanging up.

The whole neighborhood seemed to be changing, as my Mother's age bracket was aging and more young families were moving in. That was the way life intended it, I supposed.

I came home exhausted. Sammy and Ella were playing on the sun porch. It was an easy area where Sammy could crawl and pull himself up to things.

"Ready for a break, Auntie Ella?" I asked when she saw me join them. "Hey, Sammy boy, Mommy's home!" His face beamed as always, and he came crawling my way. I think I just rescued them both, which was a good feeling.

# CHAPTER 24

The next morning I walked over to check on my potting shed before visiting with Kip over at his office. I missed my regular visits to the shed, but now my plants were living in a well heated green house. There was still light snow on the ground, and I realized that I should have put on boots instead of my walking shoes.

When I opened the door, it didn't seem as frigid as I had expected. My clay pots were still there on the shelf, and tools still displayed as if I were going to use them at any moment. I wanted things to stay put as much as possible because my gardens and yard work would still require attention come spring.

I had so many memories of this shed. I cried here, wrote here, and talked to Grandmother's spirit here. This potting shed is what made me agree to purchase 333 Lincoln when Sam and I looked at the property. He loved the idea of being able to play Mr. Fix-It when he saw what all had to be done. Truth be told, he did very little of it himself, but he loved it and could see the potential. It was the right decision at the time, and I had to develop it all on my own. I hoped he was watching me as my heart still melted at its primitive beauty. I supposed it was my woman cave. I closed the door and told myself I would be back soon.

When I arrived down the road at Kip's office, I found Kevin, Kip, and Beverly in the main room around the stove drinking coffee. They appeared to be discussing the day's work.

"Hey, good morning, Anne," Kevin greeted me with surprise. "How about a hot cup of my delicious coffee?" Beverly and Kip started to laugh.

"Since I'm freezing, I'll take you up on that, but somehow I don't recall it being such great coffee!" I teased in my shivering voice. "I don't suppose you've improved that black tar, as we used to call it, have you?" We all laughed.

"Did you see we got all the geraniums in yesterday?" Beverly asked handing me my coffee.

"Yes, I did on my way over here," I reported as I sipped the coffee. "They are mighty small, so I hope they make it. You all did a great job. Mrs. Brody would also get a kick out of the Christmas lights on the barn! I love that idea so much. This whole property and my house look so totally different than ever before. You all are wonderful! I don't know what I'd do without you!"

"That's progress," quipped Kip. "The lights look extra cool with the snow I think."

"I took quite a few pictures, Anne," Beverly shared.

"Oh great, I'd love you to share them with me," I noted. "I can see using some for promotional purposes."

"How's Sammy doing today?" asked Kevin.

"He was eating more than his fair share of breakfast when I left him, but he'll work it off the way he's moving around," I bragged.

"I've been thinking about your party, Anne, and I was just discussing with these guys how we'll be able to accommodate all the cars. You never know what the weather will be like so we can't have them walking a distance to the house." He paused as if he had it figured out. "I think we'll have to get a valet service that drops them off at the south porch, and then parks the cars over here on the barn lot." He looked at me for my thoughts.

"You're right," I said picturing the event. "Do you know someone who could do that for us? I want all of you at the party having fun, not working."

"I'll figure something out," Kevin said nodding. "Can you think of anything else?"

"Well, Beverly has the house and property all decorated," I reported. "The caterer is booked, with servers I might add, and you just figured out the parking, so we should be good."

"Music, Anne," Kip nearly shouted. "You have to have music. It's a Christmas party, right?" Beverly clapped in favor of the idea.

"I suppose you're right," I said, looking at Kip, who I knew was quite the singer.

"Don't look at me, Mrs. Dickson," He said quickly. "I can handle camp fires and even a BBQ or two, but not spiffy occasions like yours." We chuckled, knowing what he meant.

"There are always musical groups that advertise on the walls in some of the buildings where I take night classes," Beverly recalled. "I'll check them out. What kind of group or music do you prefer?"

"You're in charge, Beverly, so see what's available," I instructed. "If we can't come up with something, I can have some of the street carolers stop by to sing."

"Not so cool!" said Kevin. The others laughed at the thought of it.

"Okay, Kip, something tells me you know how to throw a party," I teased.

"No dancing, Mrs. Dickson?" Beverly asked with a teasing smile.

"I suppose there should be nothing that stops that from happening," I considered. "Perhaps we can find a group that can mix it up a bit." Now they affirmed my suggestion.

"Abbey and I love to dance," shared Kevin.

"Oh, do you now?" I grinned at the thought. "Sam and I never danced much, but I can do a mean happy dance that I do at the shop as some of you may recall." Kevin doubled over in laughter.

"Alright then!" Beverly added. "I'll get started!"

I left them joking around and headed back home. I smiled to myself realizing that I had the best resources for my village. Dancing? Hmmm.

# CHAPTER 25

Ella was happy to stay home and watch one of her favorite TV shows while I went over to Sue's to babysit Mia and Eli.

When I arrived at Sue's, Eli had already been put down for the night. I was disappointed that I didn't have time to spend with him since he would prepare me for what might be ahead with Sammy. Mia was in her cute flannel pajamas that had pink cats printed on them. She loved cats, but Sue had put off having a cat because of their little dog Muffin, who was a handful. Sue told me that she may surprise her with a cat this Christmas. Mia and Muffin both were happy for my attention. Sammy was always more taken with Muffin than Mia or Eli. A dog for Sammy was likely to come in the future, I thought to myself.

Sue looked stunning in a royal blue dress that complimented her blue eyes and blonde hair.

"So, do you know where you're going?" I asked after giving her my compliments.

"We are going to Rascino's for dinner which is all I have committed to. However, he mentioned that we may also want to see a movie afterward. I don't think a long first date is a good idea." I nodded and laughed.

"Well, be as long as you like," I encouraged. "I brought a book to read after Mia is in bed. Please do not call me to check on things. You need to give him your full attention. I will call you if there is any problem whatsoever, okay? So are you nervous?"

"Very," she boldly stated. "I just hope he isn't over anxious for a relationship just because I agreed to go to dinner with him," she said with concern.

"You can control some of that, you know," I said giving her confidence. "Keep the conversation in the moment, like about the food you're tasting or the weather outside. Find out what he knows about Colebridge which should take a bit of time!" Sue looked at me like I was crazy.

"I will keep that in mind," she said with some hesitation. "You are so wise, Mrs. Dickson!"

"Well, you can tell him you're looking for a father for Mia and Eli; that should test the waters one way or another!"

She laughed. "You're nuts!" The doorbell rang.

"Hey, Devin," she said as he came in the door. "I want you to meet my cousin and friend, Anne Dickson. Anne, this is my friend Devin Rhodes." I got out of my chair to shake his hand.

"Pleased to meet you, Anne, I have heard so much about you and your lovely shop, by the way," he politely responded. "Thanks for offering to stay here this evening for us." *Hmmm, already referring to us.*

"I recall you already meeting Miss Mia here," Sue said giving Mia a hug.

"Nice to see you again, young lady," he said, trying to shake her hand. She pulled it back, not giving him her affection.

As they chatted between the two of them, I observed the average looking man that looked older than Sue, but very refined. I could sense his charm which was probably what she admired most.

Mia sensed that her Mother was leaving and hung on to her skirt tail as if she could not be seen.

"Anne will stay with you honey. You be good and do what she says," instructed Sue. Mia looked at me with hesitance.

I went over and took Mia's hand and said that we were going to find some cookies in the kitchen. She proceeded to come with me, but turned back to give her Mother a very sad, puckered up face, as if she wasn't sure she wanted to cry or go see about those cookies.

At bedtime, we went to Mia's bedroom where I read her a story that she had picked out herself. It was like magic as she fell sound asleep with the first paragraph. *Sue has her well trained,* I thought. I covered her up and went to find my book. When I got comfortable on the couch, my thoughts went to Sue and Devin.

Were they now talking about the food or were they flirting themselves into a new relationship? Sue was certainly ready for one as I don't remember her having any serious relationship. I knew she was attracted to a married man where she used to work, but she decided, wisely, not to pursue it.

It was easy for me to discard any new relationship because we had had the very best that life could offer. No one could replace Sam so I didn't think I would ever be lonely. How could anyone just settle for someone? I did want the very best for Sue as she truly deserved it. She soon found out that even having two children to fill your life, isn't quite enough. Everyone wants one-on-one attention and love. I wanted her to find this but wasn't sure Devin was the one.

I realized that I hadn't turned one page. I was staring and thinking. While my mind was on men, I was wondering if Jack Fletcher was trying to start a relationship with me by asking me to lunch. Even though I quickly declined the offer, was I flattered with him just asking? As I had told Nancy, he was candy for the eyes. He has such a clean-cut appearance. He was probably in a relationship and was just being friendly. Would he ask me to lunch if he was in a relationship with someone? I wasn't going to go into hiding after Sam's death. I knew enough to know that wasn't healthy. Dickson Properties was not going to consume my entire life. I was a woman, and I wasn't dead yet, but going out on an actual date at this point in my life was not going to happen.

Perhaps I should be planning some kind of change to make my life move forward. Maybe Sammy and I should plan a trip? I had a trip to England on my bucket list; however, I couldn't quite subject Sammy to all of that. Jean would be more than happy to go with me and would serve as an excellent tour guide. Sam would want me to go. A trip to England was the very first Christmas gift he gave me after we were married. His father became seriously ill, and we put it off. It would all happen in good time, I told myself as I leaned my head back to doze off.

# CHAPTER 26

It was almost midnight when I jumped at the sound of Sue's return home. To my surprise she came in alone but had a contented smile on her face. She sat down next to me on the couch and asked if everything was alright.

"Perfect," I said with a yawn. "It was nice, getting a little extra shut eye! Did you have a good time?"

"I did, Anne, to my surprise," she confessed with a smile. "He's an interesting man and has certainly done a lot of things in his life. Mine is pretty dull in comparison."

"I bet he didn't adopt two children from Honduras," I joked. "Did he talk much about his wife like you thought he might?"

"Not one mention of her," she said shaking her head. "I have a feeling he told himself not to. We did talk a lot about food, which is what you suggested. It turns out he loves to cook but lost interest in it just cooking for himself. Of course, without thinking, I offered to eat his cooking." She laughed at herself.

"Well, that should have opened the door for him, didn't it?" I responded.

"It probably did, but I meant it," Sue admitted sadly. "No one ever cooks for me, Anne. You don't know how lucky you were with Sam."

"You bring up an excellent point, Sue," I said, now feeling sorry for her. "So did he offer to cook for you?"

"Yes, but he didn't push it, which was good," she noted. "He just said he would keep that in mind."

"Very good," I said with approval. "Did he ask you out again?"

"Yes, we are going to see the movie that we thought about going to tonight, but didn't," she revealed.

"So you didn't see a movie after dinner?" I asked confused.

"No, we enjoyed a nice long dinner and then walked around that area of development," she explained. "I can't believe what is happening there. They are now adding a movie theater next to the restaurant. Did you know that? More condos have gone up since I've been there too! We got pretty cold, as you can imagine, so then we decided to go to Starbucks for some hot chocolate." She smiled as if she was reliving it all again.

"It sounds like the most perfect date Sue," I said.

"It really was, Anne," she agreed. "He is one great guy. Thank you so much for staying tonight. I don't think I thought of my little guys once tonight knowing they were with you."

"I was glad to do it," I said with satisfaction. "I must be going." I got up to stretch.

As I drove home, I was almost as smitten with this guy as she was. I could picture the entire night as she described it and no one deserved it more than Sue. I forgot to ask if he kissed her. Well, I would consider it so, as Jean would say.

Because I was concentrating on Sue's evening, my car automatically took me to Main Street by mistake as if I were going to work. I didn't mind since I relished seeing all the white Christmas lights which they kept on all night. I went past the depot and then the Mistletoe Market when the lights went out right before my very eyes. This can't be what I just saw so I pulled the car over to the side of the road to look again. Yes, they were totally out. I made a U-turn and slowly pulled up along the park, closer to the tent. What should I do? Was someone in the tent right now? Perhaps I should get out and see.

It was then that I saw a slender body, dressed in black, fly out of the tent. In a few seconds this person was behind the depot where I couldn't see them. It was a slim body with a black coat and sock cap, but the legs seemed to be slender looking without pants, more like black stockings. It definitely wasn't an old, heavy person; that I could be certain of. With this person's speed and shape, I guessed it was someone young.

I slowly drove further down the road to see if I could spot the person again. Was this person the one to cut the lights? What else would a person be doing there this time of night? What could I really do anyway? I couldn't be running around in the dark this time of night. I would become the suspicious one! After sitting there staring into the darkness, I finally decided to drive on home. Perhaps the next morning, Abbey could explain the activity.

I quietly made my way up the stairs and stopped in Sammy's room to kiss him on the cheek. When I slid between my sheets, my mind intermingled with Sue's evening and the mean wire cutter. I decided my twisted memories could wait until morning as I faded into sleep.

# CHAPTER 27

When I arrived at the shop the next morning, my first task was to tell Abbey what I had witnessed last night. I asked to see her in my office. I first explained my late night drive down Main Street and then what I saw happen before my very eyes. She seemed to be more concerned about my safety and was glad I did not get out of the car to pursue whoever it was.

"I guess we need to do more than what we're doing," Abbey said with frustration. "This is getting so expensive and someone, I'm sure, is getting a kick out of this. It really makes me mad. I better go down now and check on everything, Anne. Hopefully it's the just the lights and no other damage. I hope you don't mind me leaving for a bit. Sounds like Kevin will have to string more lights before this evening."

"Sure, you go right ahead," I insisted. "You know Abbey, the more I visualize what I saw; I wouldn't be surprised if it was a female." Abbey just shook her head in disgust as she went out the door.

I was in the middle of giving David his instructions for his next delivery when Mother called me on my cell to give me the details of Mrs. Carter's funeral service.

"Harry and I would really like for you to go with us, Anne," Mother pleaded. "They are having a luncheon afterward at the Barrister Fellowship room, but we don't feel we need to go to that."

"Sure, I'll pick you up before ten o'clock," I agreed. "Yes, I too, cannot take the time for the luncheon."

As I hung up, I realized how more and more Mother and Harry were becoming dependent on me. Perhaps it was just Mother wanting to see me more instead of needing help to handle Harry? Hmmm.

"Have you finished your Christmas shopping, Miss Anne?" Jean asked me out of the blue.

"I'm a pretty generous gift card person," I told her. "I thought I might visit a few shops this evening after we close since they are all open tonight. There is something extra special when you Christmas shop among the lights, smells, and music of Christmas on this street."

"Ahhh, right so, Miss Anne," Jean said grinning.

About an hour later, Abbey returned with a very unpleasant look on her face. As she took off her coat, she described that she had a messy scene of vandalism when she arrived.

"The weird thing, Anne, is I am pretty sure nothing was taken, just knocked down or pushed on the ground. There was one broken piece, but that's all. This person must be nuts is all I can say. Perhaps they are angry! What are they angry about, and why take it out on us?" I felt bad for her as she worked so hard to make this Market a success.

"Are you working tonight, by chance?" I asked her.

"No, I am here till five and then Kevin is taking me to dinner," she said smiling. "We've had very little time together with all this Christmas activity. I am also at a loss as to what to give him for Christmas! If you have any ideas, let me know."

"That's fine, dating does take some time in your life," I said, feeling pleased for her. "I am going to shop a bit on the street after work and thought perhaps we could have a bite to eat. You are such a creative person, Abbey, so I'm sure you can come up with something for Kevin. He seems like a guy who is very easy to please." She grinned with approval.

More and more the two were becoming a solid pair. I could see that it was good for both of them. I continued to work after they all left at the end of the day. I had to make sure we had things covered for the next morning since I would be at the funeral.

After I locked the door, I finally hit the bitter cold outdoors and hoped I wouldn't change my mind and head home. As I thought about the people on my list, I heard my Mother, the librarian at heart, saying you should consider "Something you want, something you need, something to share, and something to read."

I wanted to get personal, initialed note cards at Aunt Julia's shop so I stopped to see her first. She knew I was Christmas shopping. She also talked me into a darling pendant of Jane Austen to give to Jean and a new cell phone cover for myself. She was having a profitable evening which was good to hear.

My next stop was the Button Shop which was one of my favorites on the street. I really liked Barbara, who was very, very busy with customers. While I waited, I chose a button necklace for Sarah, as well as a cute little shoulder bag embellished with buttons. Artsy Sarah loved things like this. As I waited to check out, I couldn't help but admire the walls and walls of buttons. They were truly miniature works of art.

Walking out of the Button Shop with my purchases, I saw a handsome group of carolers singing across the street. About six young folks in Victorian costumes were singing God Rest Ye Merry Gentlemen, which I loved. The inspiration was very moving and I just wanted to listen to them all evening, except that I was freezing cold.

All of a sudden the Town Crier came along the street yelling that the chestnuts were being roasted up the street and that the Christmas Magician was performing magic on the steps of The Soup Tureen Shop. It was a fun, old fashioned way of letting the shoppers know what was happening up and down the street. The children loved it as he was cleverly dressed and yelled with humor in his voice.

A crowd also gathered around the many International Santas when they saw one. They each had such ethnic personalities of their own. Each Santa would provide a different card from their country for the children to collect. Albums were for sale in the Market to display the cards. This was a wonderful way to get interaction between the characters and shoppers. The Christmas committee would try to add something new each year to get folks to come back again. This year it was the Mistletoe Market.

So much thought had gone into Christmas In Colebridge through the years. It just kept getting better and better. When you talked to the visitors this time of year, each person had their favorite part they wanted to tell you about. My favorite Santa was the Americana Santa all dressed in red, white, and blue.

# CHAPTER 28

My very last stop on the street was going to be at Phil's shop. He had invited me to share a hot toddy with him and Sharon. Right now, that sounded very good. I really enjoyed Notto's hot chocolate, but I could get that across the street anytime. Nick did have a clever Gingerbread House contest going on in his shop, so I was anxious to see the entries. He had very clever ways to get folks into his shop when he wasn't griping.

I hoped that, eventually, I would get to all of the shops because they had their own way of celebrating the Christmas season with their special merchandise. Due to the cold, I decided their shop would be my last stop. Sharon and Phil were finishing up with a customer, wrapping a large print they had just purchased.

"Merry Christmas, Anne!" Phil said happily.

"The same to you guys," I responded, as I looked around to see his clever Christmas décor.

"How about that cup of hot toddy, Anne Dickson?" Sharon offered. "We've been sipping it all evening and having a pretty jolly night, I might say." She and Phil laughed as if they had a secret.

"I will. You cannot serve it fast enough! I am freezing cold." I put down my packages and rubbed my hands together to get them warm.

"Looks like you had a productive evening from the look of all your bags," Phil noted.

"I did, and it was such fun," I beamed. "If it weren't so cold out, I would have liked to have sat on a bench and listened more to the carolers. They are really wonderful."

"You know, they have to try out for that, so they probably get some really good vocalists," Sharon added. "They used to depend on volunteer church groups, but that proved to be unreliable, so money talks, I guess. It really does set the tone and background for the street when shopping. They sure earned their pennies tonight in this cold weather."

"Hey, we have a little gift for you, Anne," Phil said as he pulled a wrapped package from under the counter. "We hope you'll like it." They both chuckled. Was it from their toddy or were they about to play a joke on me?

I opened a lovely red wrapped package and commented that I did not get them anything.

"You have given every shop owner on the street a poinsettia, which was very generous, I might add," Phil praised.

To my surprise it was a framed photograph of me standing in front of the Ghostly raffle quilt we had auctioned off last

119

year at the quilt show. I had it hung on the outside wall of my shop for the outdoor show. It was still on my wall when I left to rush Nancy to the hospital to have her twins. They did the drawing and mailed it off to the lucky, or unlucky, winner without me. We had our problems with the quilt for sure. We were all glad to ship it off to the lucky winner.

"It's a little thank you for giving us a very profitable day on the street with that quilt show in the dead of winter," Phil reminded.

"Getting it mailed off to the new owner was reward enough, if you recall," I joked. "Have you heard anything from the winner, by chance?"

"No, thank goodness," Phil answered in relief. "I'm glad we took photos of it because Sharon and I put a lot of work into it. The merchants group made a good chunk with that raffle."

"I will put this on my desk at work or home and hopefully it will stay put and do me no harm," I quipped. "Thank you guys, this is very thoughtful."

We sat and chatted about business and the rumors of shoplifting as I sipped my delicious toddy. They too had a few things missing, which they were writing off as part of the season's hazard. I told them about me witnessing the runaway wire cutter last night.

"Thank goodness you didn't go after him, Anne," Sharon said.

"Or her," I snapped. "I just sense it was a female for some reason."

"I wish I had the time to catch that monster, but we are just too busy to be helpful to the market," Phil explained. "I do hope the merchants find this to be profitable. I really don't think it has affected the shop's sales on the street like they were afraid of. I think it was a new clever attraction. Many folks asked me what it was about, so many went down there because they were curious."

"I agree with you," I said. "Thanks for the present you guys, but I need to be getting on home." I got my gloves on to leave. "I'm sorry the merchants didn't have their own party this year, but there was no time, I suppose."

"George said we'd have a Happy New Year party at some point," Phil shared. "He's already dreading the next merchant meeting. He's pretty sure some are unhappy with the Market."

I gave them each a hug, wrapped my scarf around my neck, and made my way down the street in the bitter cold. When I approached my flower shop, I marveled at its Christmas beauty. The many white lights and my festive flower boxes looked most attractive with all the red bows and holly. I had one of the street's gas lights near my front door and it, too, was wrapped in the pine roping and red bows. I would be sad when all this had to go away until next year. I pulled my phone out to take a photo. This Kodak moment would be a treasure in the Christmas life of Brown's Botanical.

# CHAPTER 29

The next morning I took my time at breakfast playing with Sammy while Ella started laundry. He was enjoying one of Ella's homemade cookies she claimed were healthy for him. As we played, I had a tendency to sing or say silly poems. He sometimes laughed, but mostly he gave me strange looks. Today he grinned at my song I called COOKIES FOR SAMMY.

One cookie for my Sammy, one cookie for your Mammy
One cookie for your Grammy, one cookie for your Nanny
Another cookie for my Sammy, But that is all says Mammy
I agree, says Grammy, Then off we go, says Nanny

Ella shook her head in laughter. She enjoyed it more than he did, but I would bet he would remember it, if I kept repeating it. He was quite attached to Ella which was no surprise. Truth be told, she was the one raising him, but I hated to admit it. He called her La La and watched every move she made. He was used to me leaving but not her.

I took him upstairs with me to get dressed for the Carter funeral. He played with my smart phone for a while and then he remembered Ella as he quickly crawled toward the hall. When I handed him over to her to leave, he was a happy camper. Was this a reflection of my motherhood and crazy lifestyle?

Harry and Mother were already wearing their coats when I went to get them at the door. I noticed unfamiliar cars in the neighborhood that were likely relatives of the Carters. What would poor Mr. Carter do now without his wife?

We sat near the back of the church, despite a low turnout. I was reading the nice pamphlet they passed out about Mrs. Carter's life when I noticed at the bottom they had listed, "Officiated by the Rev. Jack Fletcher." I was about to ask Mother about the name when who did I see lead the funeral procession? It was indeed the Jack Fletcher from the Wine Gala. Was he the reverend in my church?

"Mother, who is the preacher walking the Carter family down the aisle?" I anxiously whispered.

"I told you, honey, Pastor Hamel moved, and now we have this nice young man," she explained like I should know. I was shocked and didn't know quite what to say.

"Have you met him?" I asked as she looked annoyed with me.

"Just to say hello," she whispered as she turned away to concentrate on what was being said about Mrs. Carter.

I wanted to pass out. Why hadn't Nancy introduced him as the Rev. Jack Fletcher? Why didn't he offer to tell me he was the new pastor at this church? Why didn't he offer to tell me that when he purchased the Christmas arrangements at the shop? I quickly glanced at the altar and sure enough, there sat the matched Christmas centerpieces. What was my pastor doing with a whole case of Red Devil wine?

As he so eloquently described Mrs. Carter's life, his voice was much more professional then when we talked at the Gala. We were now standing to pray, and I realized that I was praying with Jack Fletcher. I really felt embarrassed thinking back to our conversation. He was the first man to make me blush since Sam Dickson. What a fool I was!

When the brief service was over, the last thing I wanted to do was see this man face to face. The rows filed out in an orderly fashion so there was no escaping. Mother was leading the way, and Harry was behind me when we approached him.

"Mrs. Stone, good to see you," Jack greeted her as he shook her hand. "I understand you were a neighbor of Mrs. Carter?"

"Yes, I will really miss her," Mother responded. "Rev. Fletcher, I'd like for you to meet my daughter Anne Dickson." He grinned at me.

"We have actually met, Mrs. Stone," he said so matter-of-fact. Mother turned around and gave me a strange look.

"Yes, we met at the Wine Gala," I said trying to move Mother along the line of folks.

"Yes, we did," he confirmed. "Nancy and Richard Barrister are longtime friends of mine and they were nice enough to ask me to join them at their table. I was very pleased to know Richard had taken over the funeral home for his father. It's nice to have a connection when you move into a new community. Good to see you again, Anne."

"Yes, you as well," I said politely. I then looked ahead to anxiously move out the church. Who knows what was going through my Mother's mind.

We were nearly home when Mother finally asked me what I thought of Rev. Fletcher. Did I really want to tell her the truth? I could just see her reaction.

"As you said, Mother, he seems really nice," I managed to say without any emotion.

"Won't you come in and have a bite of lunch with us, Anne," Harry asked when we arrived. "I think you are a fan of your mother's potato soup, if I recall."

"Oh, perhaps a quick bite," I said to be polite. "It sounds really good, but I cannot stay long as I really need to get back to the shop."

"How about a little ham sandwich, as well?" Mother offered as she poured my soup.

"Soup will be all I need," I answered as I enjoyed my first taste. Sam loved her potato soup. How could I not think of him?

"You know, Anne, Rev. Fletcher is a single man," Harry noted out of the blue.

"What's that supposed to mean, Harry?" Mother sharply asked. Harry now looked embarrassed.

"He should be a good catch for someone, that's for sure," I added to make Harry look better. "Has anyone heard from Rev. Hamel?"

"I'm not sure," Mother answered between bites of her sandwich. "I still can't believe he's gone."

"We need new young blood in that church, honey," Harry responded. I never heard him call her honey before.

"I need to run now, Mother," I said getting up out of my chair. "Do you mind if I take Ella some of this soup? It looks like you have enough to feed an army."

"Great idea, Anne," Mother agreed as she went to get a container. "She's always thinking of others; that daughter of mine." She winked at Harry.

# CHAPTER 30

When I arrived at the shop, I stayed in my car to call Nancy. I had to know the meaning of how she handled the introduction of Jack Fletcher. She picked up the phone, and there was lots of commotion in the background. At least one of her twins was most unhappy.

"Nancy Barrister, I have a bone to pick with you!" I said boldly.

"What's happening, girlfriend?" she casually asked.

"Why didn't you tell me the Jack Fletcher at the gala was the Reverend Jack Fletcher at our very own church?" There was a pause.

"Well, for good reason," she began. "He was our guest in a very social environment. We wanted him to meet some movers and shakers in the community since he was new. It was my decision, not Jack's, by the way. If he wanted to tell folks what he did for a living, that was his business. I didn't announce that you were a florist at Brown's Botanicals, did

I?" True, she didn't. "Why are you so bothered about that?" I took a deep breath.

"Because I met him today as the reverend for the first time at my neighbor's funeral, that's why," I explained in frustration. "I nearly fell out of my seat!" I could hear her snicker in the background.

"Okay, so now you know," she curtly answered. "It was obvious that the two of you made a connection with each other, so why would you think differently of him now?"

"I don't know what I think," I said in disgust. "I don't know when you are assuming an impression was made. I just wish you would have told me about him privately."

"Get over it, Anne," Nancy said quickly. "As a matter of fact, I am going to go so far as to ask the two of you over for a dinner party I want to have for him during the holidays. If you want to refuse to come, that's fine, but I just want to make him feel at home. It's the least I can do, since we have known each other for some years."

"Okay, Nancy, I guess I am over-reacting," I admitted. "I'll have to think about your invitation."

I continued to digest all that she said after we hung up. Why did I care so much who this Jack Fletcher was anyway?

I finally went into the shop and convinced myself that none of this mattered anyway. I would refuse Nancy's dinner invitation because I would not want to send a signal that I was interested in him, because I wasn't.

I immediately started helping a customer at the counter who had chosen a Christmas flower pot to purchase. She said she had a sister who collected them. I told her there were more choices in a catalog I could show her, but she said she would look at another time. When we finished, a policeman came in the door and approached me at the counter.

He introduced himself as Captain Elsberry and told me that he was walking the street to ask shopkeepers if they'd seen any unfamiliar Christmas characters on the street. He could hardly ask me with a straight face because he knew it was a leading question. I had to laugh before responding.

"Every day, and lots of them!" I sarcastically teased. "What's the problem?"

"Are you familiar with all of the characters they have hired for the holidays?" he asked in a more serious tone.

"Not as familiar as Abbey," I responded. "She is working at the Mistletoe Market today, but she would be a good one to ask."

"We've gotten complaints from the Christmas Committee that there are at least two unauthorized characters that are browsing about," he described. "One is a woman that calls herself the 'Fairy Godmother.' She solicits men by giving out her card like the Christmas characters do. The other is dressed as a magician and seems to be pretty good at picking pockets. They engage folks in silly activities to distract them."

"Oh no, we don't need that!" I said shaking my head. That would be awful for this to come out to the public. "When were they last seen on the street?"

"Hard to say, but we do know that they like a big crowd and the darker places on the street," he revealed. "It's easier for them to blend in."

"I will keep my eye out, for sure," I said wondering if perhaps I did see them the night I was shopping. "What if you told the Town Crier character to go up and down the street yelling, 'Beware of the Magician and the Fairy Godmother, they're wanted by the Santa Police. Has anyone seen them?'"

"Hey, you're good," the Captain teased. "It might be worth a try."

"Most folks would think it fun and play along by telling you where they saw them," I encouraged him.

"Here's my card and if I may have yours, I'd appreciate it," he said before going to the door. "I will pay a visit to the Market and ask for Abbey. Thank you!" Out the door he went.

Sally and Jean were listening as they pretended to be busy.

"Skanky goons they are," Jean blurted out. "They need to be nicked straight away!"

"I agree," said Sally, trying not to laugh at Jean's reaction. "If they know the police are on to them, you'll likely not see them again. It's scary because most of their audience are children. One has to be so careful."

"How is your little tyke, Miss Anne?" Jean asked.

"He's good and on the move right now," I said with a smile. "We get him to stand alone at times. I'd give anything for Sam to see the fun things I'm experiencing right now."

"Rightly so, Miss Anne," Jean sadly said.

# CHAPTER 31

Either folks were feeling sorry for me, or I was receiving a lot of invites this holiday season. Last year I was truly in mourning and not on anyone's radar screen. This year I guess they thought I should be ready for Colebridge's social life since I was having a sizable party of my own.

One of my invites was from Uncle Jim who was having an open house this weekend. He was nice to include a hand-written line that said I was welcome to bring Sammy. I was curious to see his new digs in the Foundry Loft where Sam was living when I met him. They were pricey and overlooked the Missouri River. I wasn't sure it would be an occasion for Sammy to attend, but it was nice of him to mention it.

Despite the town's preparation for the festive season, the weather had a mind of its own tonight. It was bitter cold, but the sleet was now turning into snow and according to the weather, there could be a lot of it. This could be devastating to the merchants' right in the midst of their retail season. Many counted on this income to get them through the dark winter. I kept looking out the window and couldn't help but admire the beauty when I looked at our patio room and saw that it was turning completely white outdoors. It was Sam's favorite spot and it was especially romantic when we could light the fire pit.

The house was so still. You could hear every swish of wind and frozen drops that the weather dished out. Sammy had been asleep for hours, and Ella had turned in early as well. This big house was huge, even for the three of us. The only light I had on was from the giant Christmas tree by the stairs. It had enough light to radiate into each room, it seemed.

I went to the kitchen to make myself a hot cup of tea. I wasn't the least bit tired so I glanced at the morning newspaper that was still on the kitchen table. I couldn't help but see a very small article that had a headline reading "REV. FLETCHER TAKES OVER FOR REV. HAMEL." It said he came from a congregation in Green Bay, Wisconsin where he was an associate pastor. This would now be his first position as Head Pastor. He was called from a list of fifty potential candidates. He was quoted as saying he was pleased to be going to a community the size of Colebridge and looked forward to making it his home for some time.

I continued to sit there drinking my tea which was likely contributing to my wide eyes. It felt wonderful to sit and do nothing other than stare and recap what was taking place in my life. After the busy holidays, what would I be thinking or planning? Would I keep the status quo? Would Sammy's increased activity change things? All these questions were good to ponder, but there was something missing. Of course, it was Sam, and I did truly feel alone. Until Sammy got old enough to understand, there would not be anyone to share things with one on one. Family, friends, and employees were busy with lives of their own. My own mother was now even preoccupied with someone she cared about more than me.

I looked outside one more time before heading upstairs. I could see the snow piling up on the greenhouses that weren't even there last year. I couldn't help but hope they held the warmth and moisture that they were built to do. There were so many changes since just last year. I turned to go up the stairs and stood for a moment in front of the lighted tree to give thanks for my many blessings. Yes, God had challenged me by taking Sam, but his blessings were abundant and I knew I was very fortunate.

I checked on Sammy as he once again managed to remove his covers. He lay there without a care in the world. Suddenly, I felt warm arms around me and the aroma of Sam's after-shave. There was no question that he was with me.

"Sam," I whispered. I closed my eyes to picture him here with his arms around me. "I miss you so." I looked down at Sammy who now had a big smile on his face as he slept. Was Sam talking to him?

When the feeling left, I went to my room to try to fall asleep with my tears. Did it take very quiet moments like this to listen and feel those who had passed away? Were we usually too busy or noisy to notice when they were with us? Grandmother always used the visual of her lilies to communicate. The smell of Sam's after-shave was a special gift for me right now and told me I truly wasn't alone and that life would be okay as I moved forward. It now gave me the same smile that Sammy had as I found my sleep.

# CHAPTER 32

My to-do list was quite lengthy for the next day which was partly due to preparations for the Christmas party. I didn't want to miss Uncle Jim's open house tonight so I would need to be focused to get everything done. I was afraid to look outdoors, knowing it would be frightful. I was getting my boots on when Mother called to tell me she'd had a terrible night with Harry. She had panicked and called for an ambulance during the night. She said the paramedics were able to stabilize him so he didn't have to go to the hospital. She said that he hyperventilates when he struggles with breathing which makes it worse.

"I think he is much better this morning, but of course he shouldn't be out in this weather. If you go to Jim's house tonight, would you send our greetings and regrets?" Mother asked with a disappointed voice.

"Of course, Mother, and I am so sorry to hear this," I responded. "You are smart in keeping him away from this bitter cold air. You don't want him to be sick for Christmas."

I continued to console her as I felt her pain over the phone. Mother had reasonably good health, and she should be enjoying life instead of becoming a caregiver. Her heart was with Harry, however, and she would do anything for him. It was good that they had enjoyed some travel in the beginning of their relationship. Ella had heard my conversation and expressed her concern.

"You're right, Anne, Harry needs to stay in until this weather breaks. That goes for Sammy, too," Ella insisted. He's been a lucky fella not to have come down with anything this winter."

"Yes, ma'am," I answered. "I, on the other hand, need to first check in at the shop before I stop at the caterer's. Hopefully, the hill has been treated. Beverly is expecting me to get with her today on approving the CD of the music group she found for our party." Ella shook her head at me like I was crazy trying to get all these things done.

When I made my way into the garage, I noticed that our drive had not been cleared as it usually is in bad weather. I went ahead and started my SUV and drove as far as Kip's office. There were no cars or trucks anywhere, which was unusual. There was no evidence of tire tracks, and the snow kept coming down. This told me that my crew could not make it up our hill, and I wasn't brave enough to go down the hill in fear of not being able to stop at the bottom, or even worse, getting stuck. I turned around and parked in the garage in frustration.

After I came in and took off my coat, Ella wanted to say that she told me so but instead gave me a little grin. I picked up the phone to call the shop and was pleased when Sally answered.

"I'm sorry, Sally, but I'm stuck here on the hill, I'm afraid," I said with frustration mounting. "What about the wedding delivery today?"

"We're okay," she stated. "David made it in early, and he's back safely. Thank goodness it was only over at St. Peter's Church. I feel sorry for those poor souls today. It's horrible out there."

"Well, that's good," I said feeling better. "Call me if you need anything. Street traffic will be dead, so if you need to, just close for the day."

"I have plenty to do here, and there's no point in leaving until they get the roads in better shape," she explained. "Do you think Jim will cancel his party tonight?"

"I haven't heard anything, but if they don't come to clear my drive, I won't be going, that's for sure," I stated.

When I joined Ella and Sammy in the kitchen to get a cup of coffee, she announced, "I think it's the perfect day to make some Christmas cookies, don't you Mr. Sammy? We have lots of occasions coming up, and I think I can even get Sammy here to help us."

"He can?" I asked laughing.

"He sure can," Ella said with a snicker. "You get your camera while I mix up a batch of dough. Sammy's got two Grandma's and one Grandpa that would love to have a cookie ornament with his name on it, and if we have some luck, it'll include Sammy's little hand print."

"Oh, Ella, that would be awesome," I said sounding like a little kid myself. "You are so creative and thoughtful. He will love this!" Ella put on her-full sized Grandmother-like apron, and I grabbed my camera out of the kitchen drawer.

While she prepared for what she promised would be a big mess, I quickly went into the study to check the emails on my laptop. Like a bullet, an email from Ted jumped out.

*Hey, Anne, assume you had plans to go to Jim's party tonight. With the bad weather and all, I thought I would offer to give you a ride so you don't have to drive in all of this mess. If it suits, tell me what time you'd like to go, and I'll come by. All the best, Ted.*

Good heavens! This guy has more gumption than I ever gave him credit for. There was no way I was going to accept. I was already regretting that I told him I would have a cup of coffee with him after the first of the year. He wasn't going to get any encouragement from me in any way. After a deep breath, I responded with a "No thanks." I didn't want him to think I was afraid of the weather and the drive, nor was I going to tell him I wasn't going! Why did I find myself dodging bullets with Uncle Jim, then Ted, and even Reverend Jack Fletcher? Hmmm.

# CHAPTER 33

"Oh, no," I yelled as I came back into the kitchen. "What happened here?"

"I couldn't resist when he wanted to touch the flour, so I gave him a little to play with," Ella said, laughing. "He'll need a little flour on his hands anyway when we make the hand print." Sammy was covered in flour and looked like Casper the ghost. It was terribly funny, and he loved it.

I grabbed my camera and he played right into the drama of it all. Ella looked pretty awful herself, so I managed to get a shot of her as well. Ella was still focused on getting the cookies made. She had old cookie cutters of her own that were in Christmas motifs.

"Now take a shot of Sammy pressing down on the cutter," instructed Ella laughing. As soon as she attempted the pose, Sammy threw it on the floor which was a practice of his. Ella picked it up to start over.

"Sammy, Sammy, look at Mommy," I called as I clicked away on my camera.

"Will we be able to eat these cookies?" I asked Ella in jest.

"Here, taste this delicious dough," she said as she handed me a spoonful. "This is the only cookie recipe we ever used at our house no matter how we decorated them. My grandmother said her mother used it as well. Now, you distract Sammy while I get his handprint." She had the idea and it worked beautifully. When Sammy realized what was going on, he kept pulling back his hand.

"We have enough now," Ella said as she started to clean up. "I'll get this first sheet in the oven. We'll ice them later and put Sammy's name on them." Little childhood memories of my own were now reminding me of how I helped Mother ice the cookies. My favorite part was adding the sprinkles.

We were a mess, and so was the whole kitchen. I took Sammy upstairs to give him a bath while Ella finished up in the kitchen. We were starting a tradition today I told Sammy. His first batch of Christmas cookies was now in the oven.

Thinking of how Sam would have gotten a kick out of this, I said aloud, "Are you laughing at me, Mr. Dickson?" I felt his presence somehow. "I could use a hand right now."

After this ordeal, Sammy was ready for a nap, and so was I. I took my own shower and laid across my bed for what I thought would be just a minute or so.

I was startled out of my sleep when the land line rang on my bedside table. Ella must have picked up because it stopped ringing after a couple of seconds. I jumped up to look out the window to see if the weather had improved. The snow had stopped and someone did show up to plow the driveway. He must have come while we were having the flour storm in the kitchen or when I was in the shower. I went downstairs to find Ella still baking cookies in the kitchen.

"I can't believe you're still at this!" I told her.

"When I get in the mood and the mess is already made, I just keep going.

"It really smells wonderful!" I said taking a deep breath.

"How about a chocolate chip cookie?" she asked, scooping one off the cookie sheet. "I know it's your favorite and its right out of the oven."

"Just what I needed, sweet Ella," I said as I gave her a kiss on the cheek. "I really left you a mess. Sammy and I were both exhausted after that and fell asleep."

"You both needed that," she said shaking her head in wonder. "You know, it's perfectly alright to sleep in or take a nap when you have a day off. You jump out of bed so early everyday like your life depended on it."

"No rest for the wicked, right?" I came back to say. "I guess I'll go ahead and get dressed for Uncle Jim's party. Sammy should wake up anytime so I'll keep him upstairs with me while I'm getting dressed. Why don't you lie down for a little while?"

"Not now, but rest assured, I'll be out like a light when Sammy goes down tonight," she said wearily.

Sammy was making his usual noises when I came up the stairs. I changed his diaper and then put him on my bedroom floor to play with some plastic beads of mine and his favorite caboose which he called, "choo." He seemed to love trains or maybe it was just my imagination.

Uncle Jim's party sounded pretty casual and with the horrid weather, I wanted to be warm. I pulled out my nicer black wool slacks and my cashmere white sweater which I had done for ages. I could wear my black dress boots and of course, pearl earrings, bracelet, and rings which would serve as my bling for the attire.

When we went downstairs together, we found Ella sitting with her feet propped up in the den.

"I can stay until Sammy's bedtime if you like, Ella," I offered. She looked so tired.

"Heaven's no, I'm fine," she assured me. "We'll turn on the TV now because Tyler the Train is about to come on. Sammy stops whatever he's doing when he hears Tyler's voice. I'll set him in his play station so he'll know what happens next."

I realized that Ella had Sammy's routine all figured out. It was different when I was home, and Sammy knew that too! Sure enough, Sammy clapped his hands together when he heard the noise of Tyler the Train. It was adorable of course. I loved watching him as I was being trained to be his mother.

I kissed them both good-bye, put on my heavy fur coat, and headed to a grown-up party.

# CHAPTER 34

I loved the Foundry Lofts. They were located in a very old structure that had made railroad cars many years ago. The developer chose a high-end, contemporary design that attracted professionals like my Sam. He loved living there mostly because the top floor loft overlooked the Missouri River. I knew Uncle Jim was envious of Sam's loft when he lived there.

I was relieved to find that the parking lot was cleared of snow. Uncle Jim's loft was on the second floor. His friend Harry happened to arrive at the same time so we went up together. Seeing Harry was a clue that I might be seeing other folks from Martingale's which would not be pleasant. Harry dated Aunt Julia for a short while after her divorce, but she broke it off as he became more controlling. I could hear the jovial crowd inside before Uncle Jim opened the door.

"Hey, Merry Christmas you all!" Uncle Jim said as he greeted us.

"The same to you!" I said as I gave him a hug. "Here is a bottle of my favorite red wine, Uncle Jim, I hope you like it."

"Thanks, Anne. Just put your coats in the bedroom to the right. The bar is set up in the den," he instructed.

"How have you been, Anne?" Harry asked as he took off his coat.

"Really well and very busy," I answered in my more social voice.

"I would expect nothing else," Harry responded. "I bet little Sammy is growing up quickly!"

"He is, and I think he'll be tall like his Father," I added.

I entered the study where I was pleased to find Sarah and Aunt Julia by the fireplace. I could tell they were pleased to see me as well.

"I was worried you might not come," Aunt Julia confessed. "Anne, this is Sarah's friend, Mark."

Mark gave me his hand to shake, and it was apparent that this was Sarah's date for the evening. He appeared to be clean cut and held onto Sarah's hand the entire time.

"Have you met Rose?" Aunt Julia asked with a bit of sarcasm.

"Rose who?" I asked, feeling left in the dark.

"Rose is Jim's date tonight," she said with a fake smile.

"Are you serious?" I joked. "When did this all happen?"

"I have no idea," she whispered. "Tonight's the first Sarah and I heard of her. She couldn't wait to tell me that she was his date for the evening. I think she is trying to be the 'hostess with the mostess,' if you know what I mean."

"How did that make you feel?" I asked.

"Strange, very strange," she said in a sad tone. "I just never thought he had anyone since he was hanging around Sarah and me so much."

"Yeah, I see what you mean." I nodded as Uncle Jim headed my way.

"How do you like my tree, Anne?" Uncle Jim asked as he put his arm around me. "Sarah helped me put this up. I know you like large trees so this one should be right up your alley." He laughed. "Say there's someone here I'd like you to meet. He used to work with Sam at Martingale's, but when they went under, I made sure we hired him."

"I don't know, Uncle Jim," I said with some hesitation. "I was hoping not to encounter anyone from there."

"I know, but Case Moorland is just a really nice guy," he said with affection. "He was on the board with Sam, and he liked him a lot. He's been divorced for about five years and has been very hesitant about meeting women. He noticed you right away when you walked in tonight."

"Please don't go there," I said looking him right in the eye. "I'm not interested in meeting him or anyone else from Martingale's."

It was too late. Case walked right up to where we were standing. Without hesitation he held out his hand and introduced himself. I was polite and made an appropriate holiday comment, accepting his hand. I then smiled and excused myself to get some food before he could start a conversation. Aunt Julia followed me to the buffet table.

"You really are boxed in aren't you?" she asked, looking at me like I may have just behaved rudely. "You found a comfortable place, and you're not about to let anyone in, right?" I thought for a minute before responding.

"I think you're right, Aunt Julia," I said nodding. "You seem to be pretty comfortable yourself!" She laughed.

"Well, I'm certainly not looking for a serious relationship, but I'm not a ghost either. I'm not going to turn down a nice dinner with male conversation. Frankly, I enjoy being around men much more than women. I don't have much patience for women's typical topics. They are usually fluff and have a tendency to be quite whiney." I laughed.

"Aunt Julia, you are something!" I teased. I knew what she meant, and there were times I felt the same way.

Case Moorland kept a watchful eye on me the rest of the evening. I had made my rounds and enjoyed some appetizers as I tried to avoid Rose. If I wanted to escape free and clear, I knew I had to leave soon, so I headed toward Uncle Jim to say good-bye. On the way, Case stopped me.

"Leaving so soon?" Case asked tapping me on the shoulder.

"I am," I answered directly. "I'm afraid it's past my bedtime." He laughed and nodded with a pleasant smile.

"I know what you mean," he said in agreement. "I turn in early as well. I stopped drinking some years ago and now things just don't seem as funny or gay as they used to be." I gave him an understanding smile.

"Nice to meet you, Case," I said, putting out my hand. "I hope you have a nice holiday." He was discouraged, for sure, as his smile disappeared.

"My friend here didn't scare you off, did he?" teased Uncle Jim as he approached us. His voice was slurred from his drinking, no doubt.

"Mr. Moorland has been a perfect gentleman, but I need to get on home," I explained standing between them. "Thanks for having me, and I'll see you later this week."

"You bet, sweetheart," he said, giving me a kiss on the cheek. "Isn't she a doll?"

Case was now embarrassed and politely wished me a Merry Christmas as I went out the door.

I was glad to get in my car despite the frigid cold. I looked up at the window where Sam once lived. Remembering those wonderful days was somehow a very pleasant dream that had been interrupted.

The Lincoln hill was very slick with patches of ice that remained. I slowly took my time in hopes of not sliding into a ditch. As I approached Dickson Properties and 333 Lincoln, I marveled at all the Christmas lights that were new this year. It was like arriving in heaven. All of this was where Sammy and I belonged which made me smile.

# CHAPTER 35

I had two days to finalize everything for the Dickson party. I called Mother as she was first on my list to see how they were doing. To my surprise, they had a day of shopping planned, as well as having lunch at Donna's Tea Room.

"Are you sure you don't want me to bring something?" Mother asked to be sure.

"No, we're good," I assured her. "Sounds like the weather will be decent, at least for the moment. Do you want me to have someone pick the two of you up?"

"No, we'll be fine," she said with confidence. "How's Sammy today?"

"He's under our feet most of the time," I noted. "He knows something is going on here that isn't what he sees on a regular basis. He's a smart little guy!"

"Give him a hug from his Grammy, would you?" Mother asked sweetly.

"I will," I said before hanging up.

I stopped by Kip's office to check with Beverly on the music. I took the CD with me to listen to in the car. It sounded well suited for a party. The line at the drive-through Starbucks' was long so I pulled in the parking lot so I could go inside to get my coffee and muffins for the girls.

Sitting at a table next to the line was Jack Fletcher. I couldn't ignore him.

"Good morning, Anne," he said after he lowered his newspaper.

"Good morning to you!" I said, not knowing what to really call him.

"You're welcome to join me here, if you like," he kindly offered.

"Oh, no thanks," I quickly answered. "I'm on the run, which is why I came in. The line of cars is very long today."

"Do you come here a lot?" he asked as he put his newspaper aside. "I'm afraid I got hooked on their Pike Place brand some years ago so I was pleased there was a Starbucks near the church."

"Yup, Pike Place for me as well," I said with a half a laugh.

"I have to ask you an embarrassing question, Anne," he said shaking his head. The line was moving so now I had to step aside.

"Shoot!" I replied.

"Nancy and Richard, as you know, have been quite hospitable in showing me around, and they have asked me to come with them to your Christmas party Thursday," he revealed uncomfortably. "I didn't know what to tell them as I am not a party crasher." I was stunned, and I'm sure it showed. "They have insisted that you would not mind, but I couldn't just show up without you knowing the circumstances."

"That's Nancy for you," I joked. "She is very forward, and I appreciate you giving me a heads up on that, but of course you are welcome. It's a very diverse list of folks from business, politics, and family so you would likely enjoy it."

"I'm not sure I'll be coming, but thanks for being so nice about it," he said politely. "If you need more wine, I'll be happy to contribute." I chuckled.

"Well, I've got to be on my way," I said as I moved up in the line. "Good to see you!" Why did I just say that?

He was gone when I turned around to leave the building. Wait until I get my hands on Nancy Barrister! It was not only awkward for me, but it had to be for him as well. The Barristers will turn him into a Colebridge social butterfly before you know it.

The muffins were well received when I finally made it to the shop. Before Abbey went to check on Mistletoe Market, she said that it had been three nights in a row that the lights stayed on.

"One has to be off their trolley to leg it in the snow for such a kick, I'd say," remarked Jean.

"Well, we'll see what happens tonight," Abbey wondered. "I won't be gone long, Anne."

"Oh, I almost forgot to tell you," Sally said with a suspicious look on her face. "There is a delivery for you on your desk. It's from our fierce competitor, I might add."

"Really?" I said, going to my office.

There was a small bouquet of red roses in a white milk glass container sitting on my desk. I hesitantly opened the card which read: *Had to miss Jim's party last night, but wanted to wish you a Merry Christmas. Ted.*

I had forgotten that Ted was going to Uncle Jim's party. Thank goodness for small favors that he was not there. Why on earth would he do this? My thoughts then wandered to Case Moorland who made no impression on me whatsoever. Why was I getting signals that the rest of the world was planting eligible men in my path?

As I took a sip of my coffee and opened my computer, I thought of Jack Fletcher drinking the same blend. Did I want to call Nancy and make a big deal out of her asking him to come along with them, or should I let it alone? How would I really feel if he showed up? Would a reverend even want to come to a party like mine? I wasn't too encouraging when he told me, so he probably would decide not to come. I had to admit he smelled of good soap, and his voice was so mellow when he spoke. I felt you could trust him. I supposed most church-employed folks would have those qualities.

# CHAPTER 36

I could hardly sleep all night thinking of all the details for the party. I was excited and nervous at the same time because over a hundred folks had rsvp'd to come. Who knows how many will just show up. As many mornings, I was making my mental list of what had to happen before everyone started arriving.

Ella had already taken Sammy down to breakfast so no wonder there was total silence while I slept in a little later than usual. I put on my robe and went down to join them. In his usual excitement, Sammy threw down his sippy cup to get my attention.

"Good morning to you too," I said, kissing him on the cheek.

"This is the day the Lord hath made, let us rejoice and be glad in it!" Ella surprisingly said aloud.

"That it is!" I followed.

"David came by earlier and brought over all the flowers," Ella noted. "I left them all in the foyer. They are absolutely beautiful! I wasn't sure where you had planned to put them."

"He's on the ball," I added as I poured my coffee. "He is turning out to be quite a gem."

"The caterer will be here early this afternoon, so I want to get this kitchen all cleared up," Ella said loading the dishwasher.

"I hope Sammy will cooperate tonight and go to sleep on time," I wished. "I don't want him mingling with the crowd or he'll never go to sleep."

"Frankly Anne, we need to rope off the upstairs or we'll have folks browsing around upstairs to see the house and it will be disturbing to Sammy," Ella suggested.

"I totally agree," I nodded. "Beverly will meet the music group here early when they set up. Have I forgotten anything?"

"I just hope you enjoy the whole evening, Anne," Ella pleaded. "This is quite a big deal for you, and even folks who know you would think it be a shame if you didn't let yourself enjoy it. Don't worry about Sammy because I will check on him every so often, okay?" I nodded and smiled at her concern.

"What would I do without you, Ella?" I said feeling helpless. My cell phone was now ringing. It was Nancy.

"Today's the big day!" she announced when I answered.

"Indeed!" I said with some excitement.

"Knowing you, Anne, it will be perfect," Nancy bragged. "You have the perfect house for entertaining, and I'm glad you are taking advantage of it. So here's the reason I'm calling." She paused. "We would like to bring Jack Fletcher with us tonight. It would be such a nice Christmas event for him to meet some folks, plus he'd be comfortable since the two of you have already met."

"I was wondering if you were going to tell me, Nancy," I chided. "I ran into to Jack at Starbucks and he confessed that he didn't know what to do. Of course I don't mind, but frankly I don't think he'll come."

"I know it was bold of me, Anne," she admitted. "I'll call him this morning to see if he's going or not. I hope you'll have a good time tonight, Anne." Why was that on people's minds?

"I plan to," I responded.

I placed the flowers about the house and made a final decision on where I wanted the musical group. Ella and I were busy all day and, of course, accomplished even more once Sammy went down for his nap. I turned on the gas fireplaces in the den and living room which added a nice touch. I knew the weather would be very cold and now they were talking of snow showers, which was news I was trying to ignore. I sat down for just a moment to put my feet up and gather my thoughts. Of course, my thoughts immediately went to Sam. Everything in this house and everything that went on in it was about Sam. There was that smell of Sam's aftershave again. I knew this wasn't a coincidence. I remember that the first time this smell came to me, I was in the hospital and had just delivered Sammy. Was it his way of giving me support? I couldn't help but smile. I didn't have Sam with me very long, but it was a piece of heaven when I did.

The knock on the door brought me back to reality. I peeped out through the window to see that it was Kevin.

"Is there anything you need me to do, Anne?" he kindly asked. "The guys doing the valet parking will be here an hour early. Abbey and I will be showing up later on because she has to stay at the Market until it closes."

"Thanks for everything, Kevin," I said, giving him a hug. "Did Kip tell you if he was bringing a date tonight?"

"He said he wasn't because he'd have a better time without one," he said laughing.

"Sometimes that is true," I said, smiling. "I won't have one either, and I'm sure I'll have a great time." He winked and smiled at me as he closed the door.

# CHAPTER 37

I was determined to wear something red and festive without having to buy anything new for the party. I chose a red, sheer blouse with ruffles that I had worn only once before and paired it with black velvet slacks. This was not a normal look for Anne Dickson, but this party wasn't normal either. It did call for diamonds instead of pearls so the first thing I put on was the diamond ring of Grandmothers, which Mother had passed on to me a couple of birthdays ago. I hoped she would notice and approve. I put on diamond earrings that Ted had given me before we broke up. They were stunning and not having any real attachment to them, I put them on to enjoy.

I heard the caterers working downstairs for some time, but I wanted to be all dressed for the party before the music group would arrive. When I came down the stairs, the place was all aglow, but the smell of Ella's Christmas wassail told me that it was, indeed, Christmas.

When I arrived in the kitchen, there was Ella all dressed up in a black cocktail dress and silver jewelry that I had never seen before. Sammy looked up at both of us as if he wondered what was going on and what was different.

"You look marvelous, Miss Ella," I flattered.

"Same to you, my lady Anne," she said as she bowed.

"Your wassail smells divine," I said. "I don't know if I can wait for company to come!"

## ELLA'S CHRISTMAS WASSIAL

1 Gallon apple cider

1 Large can of pineapple juice

1 Cup of orange spice herb tea

1 Tablespoon whole cloves

1 Tablespoon whole allspice

2 Cinnamon sticks

Square of muslin cloth

Small piece of string

Mix the juices together in a big pot or crockery pot. Put the spices in the middle of the small square of muslin cloth and tie the string into a little bundle. Put the spice bag in the pot and let the whole thing simmer for 4-8 hours. Enjoy!

Beverly arrived with the musicians who were all young and dressed in black. The group of four was very polite as I showed them where to set up and told them to take breaks as often as they'd like. I told them to help themselves to food and drink, but to have no alcohol while they were playing. They understood completely.

I then went to check on Michael and the caterers who were creating wonderful smells. Ella handed me a glass of wassail and some cheese and crackers.

"I know once you get socializing, you won't have time to grab a bite, so just sit a bit," she instructed me.

"Okay, if you'll join me," I insisted. She nodded and smiled.

"Here's a cookie for my boy!" she said as she then sat him in the high chair. "I think he'll be entertained with all this going on in here until bedtime."

Wouldn't you just know Harry and Mother were the first to arrive? They were concerned that the weather might turn bad so they decided to come early. They both looked quite dapper in their holiday finest. Mother also said that she wanted to see Sammy before he went to bed. He was glad to see them as well, although I reminded Mother, he was going upstairs right on schedule.

Uncle Jim and his date, Rose, were the next to arrive. I could tell Mother was not too impressed that he had decided to bring a date. Uncle Jim said he would supervise the bar area even though the caterers had supplied one. After a lot of fussing over Sammy, Ella took him upstairs to prepare for his bedtime. I reminded her to put up the rope on the way downstairs.

In no time, the flood gates opened with everyone arriving at once. I stayed near the door to greet everyone and direct them to the coat area.

"No flurries as yet," said Kip checking in with me between guests. "The parking is going well so far."

The guests were bringing me a variety of hostess gifts which was mostly wine. What else do you give a single hostess that appears to have everything? The music sounded wonderful so far with just the right level of sound.

"Merry Christmas, Anne!" said Nancy and Richard in unison. I could hardly respond when I saw that Jack Fletcher was right behind them. They all started talking to me at once in their excitement.

Jack was wearing a sharp looking suit with a red vest which told me he was confident with his wardrobe choices. I guess I never knew what men of the cloth wore outside the pulpit.

"Merry Christmas, Anne," Jack said as Richard and Nancy moved on. "You have quite a house here! Driving up the hill to all these lights was like seeing a Christmas card."

"I know, I thought the same thing last night," I agreed. "I'm so glad you decided to come. I'm sure those two will make sure you get introduced to everyone."

"Not necessary. I'll be fine," he said looking up at the big tree with a large grin.

Next in line were Aunt Julia and Sarah. I thought sure Sarah would bring her date. Aunt Julia knew what I was thinking, so I didn't say anything.

"The place looks wonderful," Aunt Julia bragged. "What a grand house for a party!"

"Thanks," I whispered in her ear as I gave her a big hug. "I'm giving you a Rose alert. She is in the study."

"Oh, great. Thanks for the warning," she said taking off her coat.

Next to arrive were the mayor and her husband, who walked in with Vicki Pointer from Pointer's Book Store. Many of these folks had never seen 333 Lincoln so it continued to be the topic of discussion. The flow continued as I greeted Sally coming in with Al and Jean. It reminded me of poor Abbey, working in the market before she could come with Kevin. The house was filling up—especially around the dining room food table. I finally felt I could leave the front door unattended. Ella approached me.

"I just checked on Sammy, and he is sound asleep," she reported with a smile.

"You didn't put anything in his sippy cup did you?" I teased. She bent over laughing.

"Did you notice Kip is going around taking pictures?" Ella added. "That is a wonderful idea."

"Leave it to him to think of everything," I said, feeling pleased.

"Oh, here come Kevin and Abbey!" I said as they walked right in the door without knocking. "So glad you both finally made it. Is everything okay on Main Street?"

"Well, we had a very busy night till the snowflakes started," Abbey revealed as she removed her coat.

"Oh no, I was hoping the snow would hold off!" I said with alarm.

"Hopefully it's not going to amount to much," Kevin said shaking flakes off his coat. "It's not sticking just yet. Hey, those guys are doing a good job with the cars out there. You must have half the town here by the looks of things, by golly!" I laughed.

The people kept coming, as I greeted Phil and Sharon from Main Street, followed by Sue and Devin.

"Welcome and a Merry Christmas to you all," I greeted them above the loud music.

Sue was beaming and dressed in a brilliant Kelly green dress with sequins. This was a look I had never seen before. Devin, like so many, didn't waste any time commenting about the house. Sue asked where she might find Aunt Julia so she could introduce Devin to her. This looked like a legitimate couple if you asked me.

# CHAPTER 38

It was strange to look up and see folks sitting and standing on my staircase with drinks in their hands as they gazed at my large tree and listened to the music. I sensed that everyone was having a good time.

"What a great party, Anne," said a voice from behind me that sounded like Jack Fletcher. He had two glasses of wine in his hands.

"I brought you some wine I thought you might enjoy," he said with a grin. "It's called something wicked like Red Devil. I noticed that you have been abstaining most of the evening, so I thought perhaps this might be a good time to enjoy a glass."

"Pretty cute," I teased. "You're probably right about abstaining, but it is a party, right?" I took the tempting glass from him. "Cheers!"

"Cheers!" he said clinking his glass with mine. This man had charm, no doubt.

"Have you tried the food?" I asked, trying to keep the conversation away from me.

"Oh my, yes," he responded. "That didn't take me long to find and taste. I couldn't help but notice the amazing kitchen you have here. You must like to cook."

"Sorry. That was designed by Sam, who was a marvelous cook," I said smiling.

"That's what Nancy told me," he confessed. "She took the liberty to show me the patio room you added. It's pretty impressive even covered with snow. She was trying to convince Richard that they needed to add one as well."

"We love it. I mean, I love it," I said stumbling over my words.

"I'm sorry I didn't get to meet Sammy," he said so politely. "He must be a sound sleeper." I grinned and nodded.

"Thank goodness," I responded. "Ella has been checking on him."

"Reverend Fletcher, nice to see you," said Sue as she approached us. "Welcome to the most beautiful house in Colebridge."

"Indeed," he said showing his handsome smile.

"I want you to meet Devin Moorland," she said blushing. Jack reached out to shake his hand and say hello. "This was a nice event for you both to meet some people."

"I'm pleased to be here, Sue," he said looking somewhat embarrassed. "I believe you and Sue are first cousins, right?" he asked.

"We are, but we're best friends as well," I told him, as I'd told Devin earlier.

"Are you enjoying a town this size?" asked Devin. "I hear you just came here from Green Bay!"

"No question," Jack answered in a more serious tone. "It's somewhat like a small town, yet close to the metropolitan area. I'm not sure I feel quite the football passion here that I did back there, but it has many interesting folks like Anne and Sue."

"I couldn't agree more," Devin responded. Jack was looking at me, and Devin was grinning at Sue.

"Well, if you all would excuse me, I am going to try some of this food that I ordered," I said leaving politely.

"Stop. Smile, Anne," Kip said as he clicked a picture of the four of us. "Now, could I get a photo of you and Reverend Fletcher over by the tree?" Kip casually asked. I wanted to kill him.

We did as he asked and Kip grinned knowing he had stepped on my toes. Just then Aunt Julia walked by giving me a strange look. When we finished, I noticed that Harry and Mother were watching as well.

"Kip, would you take one of the Reverend with Mother and Harry?" I asked, looking at Mother. She grinned with approval. I needed to get away.

I worked the whole first floor with my Red Devil glass of wine, making sure I hadn't overlooked anyone. It was great to see Ella socializing. She seemed to know quite a few of the folks. I also realized that no one was going home anytime soon. I really didn't think Harry and Mother would stay this long.

When I walked in the living room, I was surprised to see Kevin and Abbey dancing to some terrific music. Uncle Jim and Rose quickly joined them. Where was my Sam? I would have loved this moment with him. They were now joined by the Barristers who were also having a great time.

A slow, more romantic piece was now being played, and I watched more folks crowd the dance floor including Mother and Harry. This was more success than I could have ever imagined. Everyone was having a really good time.

"May I have this dance?" Kip asked me. "You know I don't have a date tonight and neither do you, so why don't we cut the rug?" I laughed.

"It's been awhile so I hope I don't step on your toes," I blushed. "I would be honored."

"So, are you pleased with everything?" Kip asked as he took my arm. "Everyone is having a wonderful time. This house was meant for entertaining, Anne." I nodded and smiled.

"I couldn't be happier, and I owe it all to you, Kevin, and Beverly," I bragged as we swayed back and forth.

"May I cut in?" asked Jack from behind my back. Oh no!

"How can I refuse a request from a man of the cloth?" Kip joked. Jack looked at me for approval.

"Don't be afraid, I actually do dance, and some say I am quite good at it," he noted. He led me to the side of the room where there was more room.

I stayed silent not knowing what to say or do. Being held by another man right now was a strange feeling, and one I hadn't had since Sam's death. We were dancing in silence, hardly moving. I noted that Uncle Jim walked by and looked puzzled at what he saw. Finally, the music stopped.

"See, that wasn't so bad, was it?" Jack said, stepping back.

"It's been awhile for me," I said, assuming he knew what I meant.

"I understand, but its Christmas, and I think we both deserve a little fun, don't you?" What did he mean by that?

"Thanks, Jack," I finally said. "Is that what I should call you?" I looked directly at him.

"Yes, my name is Jack, and I think your name is Anne. Is that correct?" I had to laugh, feeling embarrassed.

"Save a dance for me," said Richard walking by.

"I need to get some water," I said leaving Jack standing there.

I went to the kitchen to digest what just had happened. Right behind me was Nancy who couldn't wait to give me one of those looks.

"What?" I said out loud before we both burst into laughter.

"I'm not saying anything!" Nancy said with a sheepish grin. "I love that little band you hired. I'm going to remember them. Richard hasn't agreed to dance with me for some time! It was very nice! I see that the Reverend is pretty light on his feet!"

"Oh, Nancy, he cut in. What was I supposed to do?" I asked. "I shouldn't have agreed to his request."

"For heaven's sake, why not?" she quickly responded. "Single folks deserve to dance too, and there aren't that many here tonight!"

"I'm sure my Mother will have something to say about it," I added.

"Oh, stop it. Let's get back to the party!" Nancy said as she took my arm.

Kip was now dancing with Beverly. I was so pleased that she was enjoying the fruits of her labor. She and the Littletons were the only African Americans here tonight.

Sally was the first to leave which did not surprise me. She was such a serious soul and not the partying kind so I was glad she had at least decided to come. She politely said that she had a good time. It was getting close to midnight, and Mother and Harry were now getting their coats.

"We had a lovely time, Anne," Mother said. "We knew more folks here tonight then we thought we would, and the food was simply scrumptious!"

"You know how to throw a party, Anne! Did you notice that I had enough wind in my sail to dance with your Mother?" Harry teased. We all laughed.

"I did, and it pleased me very much," I said putting my arm around Harry. "I'm so glad you had a good time. Be careful going home."

I kissed them both good-bye and was relieved that no comments were made about my dance with the Reverend! Hmmm.

# CHAPTER 39

Thank you for coming, Mayor," I said as she and her husband were ready to leave.

"Are you kidding? I wasn't going to miss an opportunity to see this house," she joked. "I'd love to come back in the spring. I can tell you have lovely gardens and two greenhouses that must make a show place."

"It takes a village to do all this, as you know, Mayor, and I have a wonderful village that supports me," I bragged. "You both have a very Merry Christmas!" Behind them followed good-byes from Sue and Devin, the Kaufmans, and the Littletons. They couldn't have left merrier.

I was shutting the door against the cold when Jack approached me. "Thanks for a great evening, Anne," he said as he put on his coat. "I have to say that some of these people here tonight, I've met at church. It was nice to see them in a social setting. I'm very glad I came." I nodded without knowing how to respond. Why wasn't he with Nancy and Richard?

"I'm glad you had a good time," I said politely.

"Would you mind if I called you to have lunch one day?" he asked gently. "You have made this transition here in Colebridge very pleasant. I'd like to see more of you." I knew I was looking at him in a peculiar way.

"I don't know, Jack," I said with hesitation looking away from him.

"Think about it," he added. "I hope you and your family have a very Merry Christmas."

"We hate to leave a good party, Anne," voiced Richard joining us. "Thanks so very much. Nancy needs to get me out more often!"

"Yes, I do," said Nancy as Richard helped her with her coat.

"I'm glad you all had a good time," I said, kissing them both good-bye.

Jack gave me a wink and went out the door with them. Hmmm.

By one o'clock in the morning, the music group was packing up, and it was just the caterers, Ella, Beverly, Kip, and me left to help me close up for the evening. Kip was helping the caterers load their van which was brought to the front of the house.

"I can't believe you didn't go on to bed Ella," I said as I watched her wrapping up some left overs.

"I didn't want to miss a thing, but I'll sure be glad to kick off these blessed shoes!" she said sitting down.

"I'm going to follow Beverly home," Kip offered. "I'm worried about that car of hers. I'll check in tomorrow to see if there are any loose ends to tie up." Out the door the two of them went.

"Anything between them you think?" Ella asked me.

"No, I just think Kip is really caring like that," I explained. "They have become really good friends working on the property together."

"The dishwasher is doing the first load which means I have permission to head on up to bed," Ella announced. "Are you coming?"

"No, you go on ahead. I'll be up in a minute," I said with hesitation. "I'll turn off the lights and set the alarm."

I turned one of the kitchen chairs so that I could just sit for a minute and think. So much had happened in these last few hours. Why was I feeling happy one moment and terribly sad the next? It was Sam. Of course it was. What would Sam think if he were observing things tonight—especially me? What pain may I have caused him when I was dancing with Jack? Why was I reacting so much friendlier with Jack and not Ted or Case Moorland? Was it because he was a pastor and I trusted him? How will I respond if he really does call me to have lunch? Would that be so bad? Oh, this was such a waste of energy thinking about all this after such an exhausting evening. I took off my shoes, turned out the lights, and set the alarm before I drug myself up the stairs.

I went into Sammy's nursery to see that he was sound asleep without a care in the world. He would be my first concern forever and ever. I whispered in his ear "I love you" and kissed him on the cheek. Nothing else mattered. Nothing.

# CHAPTER 40

It was back to work the next morning with a light snow still on the ground. Ella assured me she would handle any last details to the party, but with all the help, the place was in pretty good shape. As I pulled out of the garage and drove toward Kip's office, I could still see tire marks in the field where they had parked all the cars. Kevin and Kip's trucks were already at the office, but I had no time to stop.

I drove patiently through Starbucks to get my coffee and saw Jack's car parked there. He must make this a habit I noted. I had to put him out of my mind as I had much to accomplish today.

Abbey, Jean, and Sally seemed to be beating me to the shop these days. Spending time with Sammy in the morning before I left was part of the reason. It also happened to be the reason I was taking less morning walks.

"Splendid party, Miss Anne," Jean complimented me as soon as I came in the door.

"We had a blast!" chimed in Abbey. "Would you believe some folks recognized me from the Market in the park?"

"Wonderful," I said going toward my office. I stopped for a minute. "You know Brandy Littleton is thinking about an upscale boutique on Main Street!" I shared.

"Really," Abbey said in surprise. "Isn't Mr. Littleton an alderman for the city or something like that?"

"Yes, he is," I confirmed. "They are pretty good customers of ours; I don't know if you know that. Brandy worked at Martingale's which is where I first met her. She was one of the few people Sam trusted toward the end. She helped him do some research. There are still many of those folks out of work." I shook my head in disgust. "I told her that Donna Howard had a vacancy next to The Cookie Shop if she wanted to call her."

"Do you know what kind of rent she's asking?" Abbey asked with interest.

"Why do you want to know, girl?" Sally asked in a catty fashion.

"I keep thinking about how much I would like to open a Christmas Shop here on the street," Abbey revealed. "I know I never want to do this Market again even if they decide to repeat it. If I'm going to work this hard, I want it to be for myself. However, I have learned a lot through the process."

"Yes, Kevin has mentioned that to me," I added. "You would be quite good at it. It would be a bonus to also live on the street. Maybe you could find a building that has an apartment upstairs and room for a shop on street level."

"You're crazy, Abbey," Sally said. "You see how hard Anne and Julia have to work to make a go of things. It takes quite a while to get established, too!"

"You might have a chat with Miss Julia sometime to see if she has any regrets, Jean advised.

"I really think the street would support a Christmas Shop," I encouraged. "Some shops have gotten close to it with a lot of Christmas merchandise, but there's not a dedicated shop. Being next door to a cookie shop would be a plus, that's for sure. I know Kevin would help you, but please don't steal him from Dickson Properties!" She laughed like I was joking.

"Giving Donna a call to see what kind of rent she's asking wouldn't hurt," Abbey noted. "I'll peek inside the window later."

"So living and working on the street would not be too much for you?" Sally asked in a serious tone.

"I'm doing it now, and I love it!" Abbey confessed. "I don't want to leave here!" *That was a strong statement*, I thought.

Mrs. Brighton came in to pick up her order which made us all get back to work. I went back into my office to check my emails and was delighted that many of them were thanking me for a wonderful party. Just as Aunt Julia said, written thank you notes were becoming extinct. She said some folks that come in the shop ask her what people use those note cards for. My cell rang, and Sue's name showed up on the display.

"Have you recovered from your big night?" Sue asked sarcastically.

"I have, and I'm glad it's over, but I'll miss the anticipation of it all, I'm afraid. I can now concentrate on my personal Christmas for Sammy and me."

"So what's your reading on this Jack Fletcher, Anne?" she asked, changing the subject. "He certainly had eyes for you last night, and I saw you dancing with him!" *Oh dear, here we go*, I thought.

"He's a very nice man, Sue, but I'm not ready to read anything else into it. It was Nancy's idea to bring him in the first place, and he cut in when I was dancing with Kip. What was I supposed to do? He is Reverend Fletcher for goodness sake." She giggled.

"He's not a priest, Anne," Sue reminded me. "He is a man allowed to marry and have children."

"So what's your point?" I finally asked.

"So, my point is that he is a very nice man, like you said, and he happens to have a career in ministry," she said making her point. "Have you heard him preach?"

"Just the homily for Mrs. Carter's funeral," I admitted.

"Well, he's awesome," she bragged. "You can really relate to his message. I see him over at the funeral home on occasion, too, and he is wonderful with grieving families."

"I'm sure he is," I said as I took a deep breath.

"It's been way over a year, Anne, since Sam's been gone," she reminded me. "Sam would want you to have a full life which includes more than females."

"When did you become so wise, Miss single lady?" I joked back.

"Okay, I know how strong minded you are so I rest my case," she said before hanging up.

# CHAPTER 41

"We need some help out here, Anne," Sally called from my office door.

I hung up and headed straight to the front counter where several folks were standing in line to be helped. There were two other customers browsing. Why did everyone always come at once?

"These Norman Rockwell ornaments are precious," a stout lady said.

"They are," I agreed. "We sell a lot of ornaments for being a flower shop!"

"It's a shame the Mistletoe Market is not open today," she said complaining. "I love Christmas, and it's so unfortunate we don't have a Christmas Shop here on the street like other tourist areas." I had to snicker.

"Are you listening, Abbey?" I said across the room to her. "We have someone here who is thinking about it," I told the customer.

"Really?" she said, turning around. "I would be your best customer!"

"I love Christmas as well," I told the lady as I wrapped her three ornaments in tissue. "Enjoy, and a Merry Christmas to you!"

When the lady went out the door, Abbey was about to do the shop happy dance. She couldn't believe the timing of what just took place. When I went back to my desk, Abbey followed.

"Anne, you are so nice to encourage this shop idea, but in reality, there is no way I could possibly do it," she sadly revealed. "I have no savings for any kind of capital. Kevin and I have talked about it more than once, and he's in as bad a shape financially as I am. He has been helping his brother who lost everything in the flood, so he's always strapped for money. It is a fun dream that I'm sure many people have, but right now, that's all it is. I just felt I should clarify things before they go too far."

"Oh, Abbey, I didn't mean to put you on the spot with that notion," I sympathized. "What Kevin is doing is very admirable, and he's been so dedicated to me here which I appreciate. Having enough capital is so very important. It's why so many businesses go under so quickly. There is nothing wrong with having a dream, Abbey. That's where a lot of things start, like Brown's Botanical for instance!" She looked bewildered at that remark.

"Running that market has been challenging to say the least, but I have learned so much," she shared. "I'm not sure if the Merchants Group is happy with me, however, since there's been vandalism and some theft. The sales have been pretty good, I have to say. I don't think they'll do it again next year."

Sally now came rushing to my door.

"Anne, how many ornaments did you sell to that lady that just left?" Sally asked somewhat panicked.

"Just three," I responded. "Why?"

"Well, the Norman Rockwell ornaments are now all gone, and we had quite a few," Sally reported. "We are also missing that hand-painted Norman Rockwell watering can. I know it was here yesterday."

"I didn't sell any ornaments or the watering can," jumped in Abbey.

"Jean said she didn't sell them either," Sally contributed.

"How on earth can something the size of a watering can be stolen?" I stupidly asked.

"In an overcoat," Abbey quickly said. "People with coats, especially good-sized folks, can easily slide big objects into them. We had a lot of that in New York in the shop where I worked. We were always told to be aware of people who had unbuttoned coats in the winter. One man they caught had large pockets sewn into the side of his coat." What a sight, I thought.

"Oh my word, Abbey, I'm glad you said something!" I said astonished at the thought. "I bet that's how the painting got stolen that Phil told me about. How awful! I can't believe they have hit our shop along with the others."

"Well, when he or she is successful, they continue on. Plus its Christmas which is always the worst time," Abbey added. She knew more about all this than I did!

"I hope they nick the bloody prat straight away," Jean said in her English temperament.

"We know there's a problem now so we must be more observant," I instructed.

They all went back to what they were doing as my mind wondered about how many more things might be missing. Could the nice stout lady have been a suspect? She was a good size, and she had on a coat, for sure. I don't remember if it was buttoned or not. Could she have purchased some to cover herself being a suspect? Could a watering can have been inside her coat? Goodness knows she may be back like Abbey said. Heaven help whomever if Jean discovers them!

# CHAPTER 42

It was late Saturday afternoon when I picked up some Christmas gift cards for those on my list who enjoyed them so from other years. I called Ella and told her the three of us were going out to dinner tonight, and she should pick the restaurant of her choice. She immediately suggested Pete's Pizza up the hill from the flower shop. It was a great family restaurant and she loved their spaghetti. The restaurant had been around for generations, and they were all involved in the civic and political fabric of Colebridge.

We bundled Sammy up for the cold night, and he was excited to just go bye-bye anywhere. We settled on a corner table where we felt we could go the most unnoticed. The waitress brought Sammy a youth chair to pull up to our table. He was already getting attention from folks eating near us. Thankfully he was in a chipper mood and loved the cracker we immediately presented to him. I ordered a glass of wine knowing the evening could be challenging.

"You should be out on the town with someone other than your nanny," Ella advised me with humor.

"I am perfectly happy to be with my little family," I contradicted. "Here's to our little guy, Sammy!" I lifted my wine glass, and Ella lifted her glass of ice tea. "Here, here!" Sammy smiled big!

"Maybe you should get away for a while after the holidays," Ella suggested out of the blue. "You know your Mother and I would be there for Sammy, and things should slow down at the shop somewhat. You may want to go someplace nice and warm!" I paused before responding.

"I do think it would be good for me," I said, surprising her with my answer. "My mood swings go from happy and content to absolute sadness. I miss Sam so much, and there are times I'm even angry with him for leaving me. Dickson Properties is now up and running with Kip so I'm running out of excuses why I can't leave. Sam was always trying to get me to get away, and I always had those excuses, which I now regret." We were now interrupted by the server bringing our food. Sammy's eyes lit up when he saw the spaghetti. Was there a bib big enough for what was about to happen?

"You have such good employees, Anne," Ella added as she fed Sammy a bite of her spaghetti. "Where do you think you'd like to go?"

"Good question, Ella," I said now thinking about the possibilities. I took a bite of my salad before speaking.

"You shared with me one time that Sam had given you a trip to London as a Christmas present," she reminded me. "I think it would be grand, and Sam would be happy for you to go."

"Going without him would almost be cruel," I said putting on a sad face. "I did think about going back to Door County, Wisconsin, where we had such a wonderful time, but it wouldn't be the same of course, if I went back alone."

"Why not pick a place Anne Dickson would like to go?" Ella suggested. "You and Sammy have to make new memories you know."

"I'll think about it," I said as we continued to enjoy our food.

Out of the corner of my eye, I saw Jack Fletcher having dinner with some young woman. I didn't want him to see me. This was sure a small town. He is certainly turning into a social butterfly, and his having dinner with a young lady told me that he was getting plenty of female attention.

We requested some boxes for yummy left overs and prepared to get on home. We went out another door and successfully avoided being seen by Jack.

Sammy's bath was challenged by the red tomato stains he had all over, but after a good scrub, he was ready for bed and so was I. I noted that it might be a good time to write for a change. I cherished those moments, but tonight my mind was too many places. Even a good book was not tempting tonight.

As I lay wide awake with my bedside lamp still on, I thought about what Ella had said to me. She somehow sensed my restlessness, and the more I thought about getting away, the more I embraced the idea. Should I be totally adventurous and go to England or just go to the beach and lie in the sun with an umbrella in my drink? Hmmm. I dozed off.

Sunday morning I decided to go to the late service at church so that I wouldn't run into Mother and Harry. I did want to hear Jack preach and did not want to be embarrassed about not hearing any of his sermons before I ran into him again. Ella thought my activity strange, but I avoided a conversation as I took Sammy upstairs with me to dress. It might be my imagination, but Sammy seemed to enjoy watching me go through the process of getting dressed.

I snuck in the back row as the service was about to start. I began reading the bulletin and read that Jack was teaching a Bible Study class on Wednesday nights, and another paragraph reported that Reverend Fletcher had attended the quilting ladies' luncheon where they presented him with a gift certificate for a casserole on each Wednesday of the first month that he was here. *Now that was clever for a single man,* I thought. *Sam loved casseroles, it must be a man thing.*

I had forgotten what a wonderful choir this church had produced. The selection was breathtaking and brought tears to my eyes. As I looked at them, I saw many familiar faces wearing those choir robes. Next was Jack's sermon titled: What is your Christmas gift to Christ?

He began by asking the congregation if it were better to give than to receive. The congregation quietly responded. He then quoted President John F. Kennedy: "Ask not what your country can do for you, but what can you do for your country?" He explained that it was human nature to always ask God for things we want or feel we need.

"God's grace is there for the asking no doubt, but what can we do for him? Since Christmas seems to be all about giving, what do you have for Christ on your list?" He paused before he continued. "God smiles when we share our faith with others. That would be a very nice thing to give to someone. What also makes God happy is when we open our hearts and pocket books to the needy. So many are hurting right now and cannot have the Christmas you might have this year. It could be as small a thing as a compliment or a hug! If God has blessed you, passing something on is a gift you would also be giving yourself! Remember, when you give a gift, you should give it freely with no strings attached. There may not be a thank you or a hug back. Smile inside for you have given this gift to God as well." It was excellent food for thought.

"We are blessed with many volunteers in this congregation," he continued. "I thank you and God thanks you for your unselfish gifts. If you are not one of our many volunteers here, I suggest you consider the thought. What you will receive will be a gift to yourself and a gift to God. God bless you all with a very Merry Christmas as we love and serve him. Amen."

The service continued with the collection and another hymn. I looked around the beautifully decorated church which reminded me of our church tour. It was a good message for me to hear, and Sue was right about how easily he preached from the heart. He didn't look down at any notes the whole time. I wondered if he knew growing up that he would someday be a pastor? Hmmm.

# CHAPTER 43

I left during the last hymn so that I wouldn't have to shake hands with Jack when I left the church. I really did like his sermon and would tell him so one day.

I went on home to pick up Sammy to spend some time at Mother's house. She often complained that they did not get to see him very much. Sammy was cranky the last few days with his teething so I wasn't sure whether strangers would be well received. They were thrilled to see him when I arrived, and Harry made sure they had some goodies for him to play with.

Mother nearly fell over when I told her I went to the late service by myself.

"It just worked out better for me," I explained. "I thought his sermon to be very good, didn't you?"

"Yes. He has that gift," Mother noted. "Speaking of gifts, Harry and I are at a loss as to what to give you for Christmas. Is there anything that you can't buy for yourself, Anne?" I paused.

"Perhaps there may be something," I said as I was still thinking. "I think I am going to go away somewhere after the holidays. I may go to England which was an unused gift from Sam a couple of Christmas' ago. Do you remember?"

"Oh, my Anne! Would go alone?" Mother asked in shock.

"I haven't thought that far ahead," I said in thought. "I may just go to the beach where it'll be nice and warm, but I think the change will do me good." Mother looked at me strangely.

"So helping you with a trip is what you're hinting at?" she observed. I laughed.

"Sam certainly tried in many ways," I recalled. "He even gave me all new red luggage last year for Christmas."

"If it helps financially, of course we can help you," Mother said looking confused.

"No, I guess I really didn't mean that," I explained. "I was thinking more of being helpful with Sammy."

"Of course, honey," she said still not sure of my request.

Sammy and I were both sitting on the living room floor as he safely played with Harry. Once Harry had tried to pick him up or take his hand which had negative results. As long as I was close by he was game for anything.

"I think I have something special he will like," Harry said as he showed me a paper bag. "I played with this train when I was little and hoped I'd have a grandson someday that would like it as well, but that didn't happen. When your mother told me Sammy liked trains and that she was making him a train quilt, I got this out immediately." Sammy's eyes lit up.

"Mother, you're making Sammy a train quilt?" I asked in disbelief. She looked at Harry like she was going to hit him.

"There goes my Christmas surprise for Sammy, just like that!" she fumed. "I told you not to say anything, Harry!"

"Oh Mother, how wonderful," I cheered. "I won't tell him. It will be a surprise when we actually see it. Are you finished with it?" She shook her head in disgust.

"Some folks have given me a hand with a few quilting stitches, she revealed. "Aunt Julia, Isabella, and even Sue have all been here to put a few stitches in it. Now you just act surprised Christmas Day when I give this to him, you hear?" She referred to Sammy as if he were old enough to know.

"You've been quilting in the basement without me?" I asked in disbelief. Mother finally smiled.

"I've said all I'm going to say," she stated. "You're just like you always were at five years old. You would ask questions until I helplessly gave in." Harry laughed.

"Really?" I laughed again.

"Okay ladies, I'm calling a truce here," Harry said. "Can I give this little guy a cookie?"

"No Harry, we have to get him home for his nap," I said getting up from the floor. "Feel his forehead, Mother. Do you think Sammy might have a slight temperature?" She leaned over to put her hand on his forehead and cheek.

"He might, Anne, but that is not unusual with teething," she noted. "You might want to give him a children's aspirin before his nap."

We said our good-byes, but Sammy wanted to keep the train. I told them to keep it here for him to play with Christmas Eve when we come for dinner.

Sure enough, Sammy went right to sleep for his afternoon nap without even a fuss. I was changing into my blue jeans when my cell phone rang.

"Hey, Anne, Jack Fletcher," he said as if I were waiting for the call. Why didn't he just say Jack?

"Oh, hi," I said as he caught me by surprise.

"I was sorry you rushed off in church today," he said matter of factly. How did he see me? "Was it something I said?" He was trying to be funny.

"No, not at all." Now I was feeling embarrassed. "I needed to get on home. That was a wonderful sermon, by the way."

"Glad you liked it," he said sounding more relaxed. "I'm not calling to play I spy, you know. I have a light day scheduled for tomorrow and wondered if you could swing a short lunch with me?" Here was the question, now what?

"Well, without knowing what the shop's schedule is tomorrow, I'm afraid I can't commit," I confessed.

"I promise I won't keep you long unless you have the time," he insisted. "I was thinking I'd like to go back to that restaurant, Rascino's, where we enjoyed the Wine Gala. I hear their lunches are very healthy and delicious. So how about it?" He was closing in, and I wasn't sure my defense mechanisms could handle it.

"Call me at the shop around ten-thirty tomorrow morning if you don't mind last minute notice," I suggested. "When a business is affected by births, accidents, and obituaries, one doesn't always know what it means for us in orders." How many times had I used this line?

"Sure. The same goes for me you know," he reminded me. Of course, it was the same. "An uninterrupted night's sleep is a good day for me."

That was true. At least they didn't call me at home to order flowers.

"Of course, Jack, I wasn't thinking," I said apologetically.

"We'll take this topic up in the morning then," he said as we hung up.

The plan to block him completely out of my life was not working. Having lunch at Rascino's would announce to the whole town that we were seeing each other. That would not be so smart. Perhaps tomorrow, I would have the proper solution. Hmmm.

# CHAPTER 44

The first thing on my list the next morning was to pick up my quilt from the depot. The show had concluded last weekend, and with some luck the quilt show did not have any theft occur. The risk was great because they had so many different volunteers come and go.

I stood in line with others waiting for their quilts. When it was my turn, they didn't take the quilt down until I showed them my drop-off ticket and ID. The ladies taking my information seemed to be very pleased with the attendance and thanked me for participating.

Now that the quilt was tucked away safely in my car, I went to the shop knowing I had to check the schedule before calling Jack about lunch. The more I labored over the idea, the more confused I became.

I took a call from Mother on my cell as I walked in the shop. The girls were busy and hardly acknowledged me.

"I just got a call from Sue. She won't be coming to dinner Christmas Eve," Mother stated with disappointment. I waited for her to explain. "She is spending the day with this guy Devin. Is this serious, Anne? She just met him. She said she'd like to have the holiday at home cooking for a change."

"I can understand that, Mother," I said as I defended her. "She has her own little family, and if she doesn't have to go somewhere and be a guest, it probably sounds pretty good to her after all these years. She said she'd be at my dinner so she is willing to share her time. So it'll be Aunt Julia, Sarah, Ella, Sammy and myself, I guess."

"That's right," she said sounding a bit disappointed. "I might have invited Jim, but not at the risk of him bringing that woman Rose. Not with my own sister there!"

"I'm sure he'll understand," I noted as I wanted to get off the phone. "Can I bring anything?"

"No, I have it all planned," she bragged. "It may be the last time I am able to have it here."

"Stop thinking like that, Mother," I scolded her. "I've got to go. I just arrived here at the shop."

When I hung up, I knew the sadness she felt. Everyone was going their own separate ways, and her house was no longer the centerpiece of the family like it had been for so many years.

"Anne, a man was here earlier to see you and told me to give you his card," Sally reported.

When I looked at it, I couldn't believe it. It was from Case Moorland. On the back of the card he wrote: *Would love to have a Christmas drink, call me.* Really? I don't think so. What kind of contact was this? Was this how single people communicated these days?

Before I called Jack, I looked at the delivery schedule for the day, and it appeared that Sally and Jean had everything started and under control. Abbey was scheduled to stop by later to see if we needed her. Okay, here I go.

I used the cell number that was on my phone from him yesterday. My stomach was churning as I heard it ring. Finally he picked up.

"Anne, good to hear from you," he answered. "How's the day going for you?"

"Good, I'm good," I responded. "However, I can't be gone too long."

"The same here," he noted. "I forgot we have a voter's meeting this afternoon, and I have to be there for that."

"I'll meet you there about noon then," I suggested.

"I'm looking forward to it," he said so pleasantly.

Jean brought in the mail and I was surprised by how many folks were still sending Christmas cards. I loved them, but my list was becoming shorter every year. It was still a good business practice to remember certain business contacts and out of town friends and family. I included a letter to Helen with a photo of Sammy in my Christmas card to let her know I was still hoping they would come for Christmas.

There was a Christmas card from Ted which was no surprise. He once again reminded me that he was looking forward to having lunch after Christmas. Why on earth I fell for that request, I didn't know. Perhaps I could still think of a way to get out of it. I would open a whole new can of worms if I gave him any sign of encouragement.

I cleared a very messy desk before lunch, and fortunately, no one asked where I was going when I left the shop. It was hard keeping any secrets in a small shop.

I walked in the restaurant on time, and there was Jack standing by the reception desk.

"Really glad you could make it, Anne," he greeted me with a big smile. We then followed the hostess to our table. I looked around the crowded room and so far had not seen anyone that looked familiar. When she handed me the menu, I glanced up to see Jack looking at me with a smile.

"What are you so happy about?" I casually asked.

"You. I can't believe I got you here!" he said gloating. "Besides the Barristers, I didn't know a soul here, but I feel I've hit a home run meeting you! I appreciate you taking some of your precious time. I know how busy you must be."

"You're embarrassing me," I said shaking my head.

"I guess we can't order some Red Devil in the middle of our day, can we?" he teased.

"Don't tempt me," I responded. "I'll have this seafood salad." I said pointing to the item on the menu.

"That's sounds good to me as well," Jack followed. The waitress now left us alone.

"So how is Sammy doing these days?" he asked due to a lack of a topics to choose from.

"Teething and very fussy if you really want to know," I complained with a smile.

"Has he made his visit to Santa Claus?" he asked grinning.

"No, I'm afraid we're not ready for that yet," I said with some seriousness. "I'm not willing to scare him to death just to get a good photo opportunity." He laughed nodding.

"Do you have a photo of him on your phone you could share?" he asked catching me off guard.

"Oh, sure. What Mother wouldn't?" I bragged as I got out my phone.

I first showed him the ones I took when we were making cookies and he was covered in flour. The other one was Sammy in the stroller when we were at Sam's grave site.

"He's a good looking little kid, Anne," he noted. "His father must have had dark hair, I presume."

"Yes, very dark," I added. "Sammy looks nothing like me."

We devoured our lunch and talked about a variety of topics. He certainly was easy to talk to, but then it was probably because he was a man of the cloth and heard everything under the sun.

"Do you have any time this week to take in one of the Christmas concerts at the University?" he boldly asked as he wiped his mouth.

"Afraid not, Jack," I quickly answered. "Retail is very demanding this time of year, plus I have a lot of family commitments."

"I understand," he said nodding. "I wish I could go back home to my parents for the holidays, but that's just not possible with me just getting here." I nodded in understanding.

"So do you like it here, okay?" I asked looking into his innocent face. I wanted to ask so many questions about his family and if he had ever had a serious relationship, but I left well enough alone.

"Oh, more and more, Anne Dickson," he said, as his face lit up.

We left with almost a business-like farewell and once again wished each other a happy holiday.

When I returned to the shop, Abbey was frustrated. She was on the phone trying to get volunteers to help take down the Mistletoe Market that would end Christmas Eve afternoon.

"I'll try to help you, Abbey," I said as she hung up. "I can imagine how busy everyone is, and I don't have any responsibilities that day since I'm going to Mother's for dinner."

"I could really use it, Anne," she cried hopelessly.

"I think what little business we do on that day can easily be handled by Jean and Sally unless we have some emergency orders," I explained.

"Kevin isn't available either," Abbey complained. "He has family coming in that he has to deal with."

"Surely you'll be joining him for Christmas Eve, won't you?" I bravely asked.

"Oh, yes. He made sure that I was included," she blushed.

"The girl is smitten," Jean teased. "I knew straight away Kevin was smitten with her as well, did you not also, Miss Anne?" We laughed.

"Humbug," joked Sally aloud.

"Just come on down anytime you can make it, Anne," Abbey instructed me. "I have got to run. I'll see you guys later."

With my back turned from the girls, I heard a person ask if there were any more Norman Rockwell ornaments. As I suspected from the sound of the voice, it was the same stout lady that had purchased some from me before.

"Oh, hello there," I greeted her. "I guess the ornaments were a big hit if you're purchasing more." She nodded, but I could tell she didn't really remember me.

"We have these cute little snow babies half price," Sally said as she picked one of them up to show her.

"No, no, I'm into Norman Rockwell things," she explained with frustration.

"Oh, well, then you'll want to take a look at this hand-painted flower pot that we have in the window," Sally said as she turned to retrieve it. She paused as I watched. "Well, we did have it. Jean, did you sell it?"

"No, ma'am," Jean said with certainty. *This was not good*, I thought.

"Well, I'm sorry. I guess we no longer have it," Sally apologized. "Is there anything else we can help you with today?"

"I'll just browse a bit," she shrugged.

I watched the customers thinking there was something odd about her. I pretended to be busy, but I didn't let her out of my sight. All of a sudden there was a swift movement.

"Oh, Miss, I think you dropped this," Jean said picking something up from the floor.

"Oh, I'm just trying to carry too many things," the woman said with frustration.

"I see that you are a Jane Austen fan," Jean announced looking at the Jane Austin ornament that she had picked up off the floor. The lady nodded.

"Well, I must be going," she suddenly said as she walked toward the front door. Sally quickly walked ahead to block her path.

"Don't you want to pay for that flower pot before you go?" Sally asked shocking Jean and me. The lady hesitated and I almost fainted with fear. What was Sally doing?

"Oh well, I guess I didn't realize that I must have been looking at it," she said fussing with her coat. "Of course!" She turned toward the counter.

"Jean, please lock the door," I instructed her when I saw what was happening.

"I think we have ourselves an old-fashioned shoplifter," Sally said with anger. I picked up the emergency number for the mounted police on the street.

"I knew you were the only one who could have taken that pot and goodness knows what else," Sally said now raising her voice.

I wanted to calm Sally down in fear she might be falsely accusing the lady.

"You're crazy! How dare you insult me like that," the lady responded in anger. "I was going to purchase these items, and you cannot keep me here!"

"You can explain all that to our friendly copper coming to our rescue right now," Jean told her.

I knew what our shop had done was very risky business. We had never caught anyone red-handed before. I had always been warned that you should wait to approach a shoplifter once they were off your property to prove they truly did not intend to pay for the merchandise. They could turn around and sue us for humiliation and being falsely accused. Sally was certainly aggressive, but it worked, at least for now.

Jean let in the officer, and after Sally explained the circumstances, he asked the lady to remover her coat. She became irate, but his approach made her feel that she had no choice. There was considerable merchandise inside the coat and he told her that if she had receipts for everything, it would be very helpful. Of course, she could not produce them. When I saw more Jane Austen ornaments, I told Jean to tell Aunt Julia to come over knowing they likely came from her shop. She had also lifted two very nice watches that she put in her bag of cookies from the Cookie Shop down the street. It explained how and where we had lost other things.

The officer radioed for assistance to take away the lady who was becoming more and more angry. As he took her outside the door, she abruptly said, "You'll be hearing from my lawyer, ladies!"

# CHAPTER 45

I cannot devise how she popped that pretty pot in her wrap," Jean said in shock. "I actually took a shine to her at first, but then after I saw the Austen ornament spill out, I knew straight away she was a blagger."

"Good job, Sally," I said in disbelief to it all. "You have a very good eye! Did you really know for sure she had a flower pot somewhere?"

"Not totally, but I wasn't about to let her leave with it," she said with certainty. "I just knew it had to be her and no one else."

"I thought Miss Anne was going to have kittens when you blocked the ole gal," Jean said with laughter.

Just then, Aunt Julia came in to tell us that the officer had come in so she could identify the stolen ornaments. She was so grateful and blown away that we had discovered the thief. She was furious that she had been nice to the lady and had not suspected anything.

"I would venture she has popped in more than a shop or two here on the street," Jean chimed in.

"Well, we won't get everything back, but I'm glad we stopped her," I stated as I tried to get everyone back to their tasks at hand.

It was no surprise when nosey Nick from across the street was the next to come in for information. We all explained what had happened and watched his blood pressure rise.

"I put very little merchandise on my counter these days," bragged Nick. "Everything is under glass. If I ever catch anyone, look out!"

"While you're here, Nick, I need to order your triple layer chocolate raspberry cake. I'll pick it up Christmas Eve before you close so I'll have it for Christmas day."

"I'm closing at three o'clock sharp that day, so don't forget about it," he scolded. "I wouldn't want to be stuck with that. It's a lot of work!"

That was Nick. Instead of thanking me for the order, he nearly threatened me to pick it up in a timely manner. He always had to have the last word, and it was always negative.

"You're very welcome," I teased, hoping he would get the hint. "I'll be there, so not to worry." He nodded and went out the door. The girls shook their heads in disbelief.

"I say he should play the old scrooge for the Christmas characters on Main Street, "Jean joked. We laughed.

"He is the scrooge on Main Street!" I joked back.

"I hate to bring this up, Anne, but I realized we don't have a stocking hung up here for David," Sally divulged. "Should we use Kevin's old one?"

"Oh, my, no," I responded. "I wish I would have thought about that before now. I ordered one for Sammy from that shop, but I'm a bit late to order one for David this year, I'm afraid. Perhaps Abbey can find us one at the Market and we can get his name on it eventually. By the way Sally, I'm expecting you to join us for Christmas dinner. Jean, if anything changes with the two of you, please join us, okay?"

"Much thanks, Miss Anne," Jean responded rather sadly. "In vain I have struggled to get on to my England home, but Al has no sensitivity to it at all I'm afraid. He is a bit out of sorts again with the holidays coming about."

I ignored her more frequent cries of home sickness, and I was hoping that when she said Al was out of sorts, it didn't mean he was back to drinking. He always insisted they have their holiday meals with their next door neighbors which were more to Al's liking than Jean's. I told myself that if I did pay a visit to England, I would surely take Jean along. If it was a matter of money, I would provide it. The enjoyment of her company and knowledge of the country would surely be worth it.

I left the girls to go back to my computer to email George, our Merchants President, about the shoplifter so he could alert the rest of the organization. I wondered if anyone else on the street had suspected her.

Why did every good thing have to have a bad thing attached to it? Why couldn't everyone enjoy what they could of the Christmas season instead of causing pain and loss to someone? Was the pleasure in revenge for some reason? Was the person a victim herself that she wanted to strike back? How far should our spiritual responsibility of forgiveness go when they are doing wrong? Would we be doing them a disservice by not letting them pay for their crime? Perhaps this was a good question for the new Reverend Jack Fletcher to answer!

# CHAPTER 46

After everyone had gone home, I picked up the officer's card that was left with us in case we would find anything else missing. I'm sure every business person wanted to trust their customers, but I remember Isabella telling me that the shoplifting in her quilt shop came from customers that she trusted most.

As I locked the door, I observed many of the Christmas characters still active on the street. I heard carolers singing, but I was not sure where they were standing. In a quick second, I saw the Sugar Plum Fairy dash by wearing a black coat as well as black tight stockings. It reminded me of what I saw leaving the Market tent late one late night. Surely it wouldn't be her. My instincts at the time told me it was a woman, and she fit the description. Perhaps she had a reason for being there since she was one of the street characters. By the time I opened my door again to get a better look, she was long gone. Since Abbey said that they still did not have a clue on the light cutter, I would have to pass on my suspicion. Perhaps their suspect was right in front of their eyes just like the shoplifters!

I went back to my desk to prepare inventory sheets for the end of the year. It seemed like we just completed the task. My responsibilities were greater to the IRS this year with Dickson Properties, but Kip was being very helpful and keeping good records.

I wanted to get home before Sammy's bedtime and it was getting late. As I headed toward the door to leave, I couldn't believe it when I came eye to eye with Case Moorland.

"Anne, I was hoping to catch you when I saw your car was still here," he said anxiously. His appearance surprised and scared me at the same time.

"What do you need?" I asked in a shaky voice as if he were a customer. "The shop is closed."

"I realize that, but when I saw you were here late, I thought I could talk you into that Christmas drink after a long day," he flirted.

"Oh no, Case, I have to get home before Sammy's bedtime," I said firmly as I kept moving.

"Why are you playing so hard to get?" he asked peering into my eyes. I couldn't believe what he had just said.

"I don't play hard to get, Mr. Moorland," I said with my voice rising. "You don't know me very well. I am a busy woman who means what she says, and if I were the least interested in seeing you, you would be the first to know." He was shocked by my response.

"Okay, okay, I get it, I get it," he responded loudly. "Even widows deserve a little fun now and then!" I headed straight to my car parked in front of the shop, as he asked, "Why don't you at least give me a chance?"

"Merry Christmas!" was the last thing I said to him as I got in my car. I drove away quickly as if he were going to come after me. I didn't know this guy and I had no desire to. I really resented his referral to me being a widow that might need some fun. Was Uncle Jim pushing for this relationship to happen? If I ever started dating again, which didn't sound very appealing right now, I would not be attracted to someone like Case. By the time I got close to home, I had settled down and almost felt a little guilty about my abruptness. The only good thing about this encounter would be that Mr. Moorland finally got the message.

Sammy was up and very fussy as Ella was trying to give him a bath. Normally he loved this time of day. Of course when Sammy saw me, I was the only one that could take over. Ella was happy to concede as she'd likely had a challenging day with him.

I dressed him for bed and sat him in the middle of my big bed surrounded with pillows to read to him. I didn't read but a few pages and he leaned to the side in slumber land. I carefully lifted him up to take him to his crib and noticed how his weight had continually increased. When I returned downstairs, Ella offered me a drink and a bowl of the chili she had made. Sam made the best chili ever, but Ella's would do nicely since I had no dinner. The minute I sat down, she came to me with a list of guests whom she thought may be at our house for Christmas dinner. I continued eating as she read her list.

"Besides you and me, I have Sylvia, Harry, Sue, Julia, Sarah, your Uncle Jim, Kevin, Abbey, Beverly, Kip, Sally, and hopefully, Helen and Pat. That would be eighteen. Are you sure William and Amanda won't be coming?"

"No, William told Mother they wouldn't be coming home for Christmas," I shared. "I don't know what to make of Helen and Pat. They will not commit and I know it's because of Helen's health. She probably is a little miffed that I don't come up there for Christmas."

"This is Sammy's first Christmas, and maybe you can visit her at the first of the year!" suggested Ella. "Now I did not figure in that friend of your Uncle Jim's. Rose I think her name is."

"No, she is not invited," I made clear. "This is my Aunt Julia's family and it would be pretty nervy for him to bring a date when technically he is no longer a part of our family."

"That's what I thought," Ella said in agreement.

"We probably will have to add Devin to the list," I remembered. "I think they are becoming a couple, and I have a feeling Sue will want to ask him to come with her."

"So there will be twenty people before we know it!" bragged Ella.

"The more the merrier!" I cheered.

# CHAPTER 47

I showed up at the market to see how I could be helpful with dismantling everything. When I got there, I could see why Abbey was overwhelmed. It appeared disorganized and customers were still popping in to see if they could buy something at a discounted price. Abbey said that all of the small live trees would remain in the park for the Christmas season and be displayed around the depot. Since much of the merchandise came from the shop owners, some were picking up what was left.

Abbey knew I looked helpless so she directed me to help the character, Jack Frost, take down the roping. Jack was in costume and ready to go on duty for the evening. He was a happy guy and seemed pleased to be helpful to Abbey. He climbed the ladder and I caught the greenery as it fell to the ground and put it in the trash bags. He was a talkative young man so I asked him if he knew the Sugar Plum Fairy very well. He laughed.

"She's a strange one," he claimed with laughter. "She knows she won't be back next year so she's driving everyone crazy. I avoid her."

"Why?" I asked curiously.

"She's trouble, that's why," he said in a more serious tone. He took my trash bags to the truck outside the tent.

I returned to Abbey to get my next assignment.

"These lights up here and on those poles are still good so we need to save them for the Merchants Group to reuse next year somewhere," Abbey explained as she pointed. "Goodness knows they paid for plenty of them. Most of them you should be able to take down without a ladder. Jack can help you if you need to use a ladder."

"Got it," I responded. "Oh, Abbey, I wanted to ask you if there were any leftover Christmas stockings. I need one for David."

"Sorry, we don't have any," Abbey reported. "Hey, I'm happy to share mine at the shop." She grinned.

"No, he can share Kevin's stocking since Kevin's not there anymore," I noted.

"Good idea, Anne," Abbey agreed. "Look at all these boxes of tinsel we have left." Abbey held up a few boxes to show me. "I don't think we sold one! I guess no one uses tinsel anymore, but I thought it old-fashioned enough to be perfect for this market. I remember my Grandmother putting tinsel on her tree one strand at a time. It took forever. We could help her with the ornaments but not the tinsel. It was her final touch!"

"Oh, when I was younger, we put tinsel on the tree but not strand by strand," I recalled. "I'd get in trouble when I'd play around and just throw it on I do remember that. I do love the look of it but would never put it on a big tree like mine because it is quite messy as the days go along. We used to find strands of tinsel left around the house days after the tree was taken down."

"Well, I guess I'll pitch this," Abbey said with frustration.

"No, please don't do that!" I said surprising her. "Somebody would love to have that!"

"Everything has to go, Anne," Abbey claimed. "I am getting paid to get this all down and cleared away." It was at that moment that I had a brilliant idea.

"I tell you what, I will purchase anything you have left at half price," I offered. "I have room to store it all in one of the former bedrooms at the Dickson office. I'll keep things there till someone decides to open a Christmas shop one day!" I thought Abbey was going to faint right in front of me.

"No way, Anne," she shouted across the tent.

"Deal or no deal?" I asked with my hands on my hip. "The merchants need the cash right now, and they may even want it for another Market next year if someone doesn't open a Christmas shop."

"That would be very generous of you, Anne," she finally admitted. "A lot of this stuff was donated by the merchants and they don't want it back."

"Well, so put all that stuff on a pile and present me with a bill," I instructed her.

"Okay, if you're certain!" she said as she shook her head in disbelief.

We kept on working, and I got dirtier and dirtier. Finally George, Phil, and Kevin came by to help. Abbey was thrilled that we could finally finish things up. I instructed Kevin to put all of my half-price merchandise in his pickup truck and take it to the office. His questionable look was priceless, but he did as he was told. At that point, George suggested we all get some refreshments and a pizza up the hill at Pete's Pizza. It sounded wonderful, and I was famished.

I felt I deserved the treat so I called home to tell Ella that I would be home before Sammy's bedtime. So like her, she told me to have fun and not to worry. I drove there to meet them, and it was good to share a few laughs and Christmas gossip from the street.

# CHAPTER 48

It was the day before Christmas Eve, and we had an unusually full day, which now included orders from a horrific traffic accident that had happened the night before. A young mother and two children, who were two and four, were killed when a tractor trailer truck hit their car. The father and the truck driver were in critical condition. The Forester family was well known in town because generations had been in the construction business, but I didn't know any of them personally. The Barrister Funeral Home told us that the funeral would likely be held in a few days which was helpful in planning the arrangements we would have to make.

Mother called me first thing to tell me that the story had been on TV. She said she knew that the mother's parents and whole family had frequented the book store when she worked there. As busy as I was, I felt her sadness and her need to talk.

"Speaking of the book store, Anne, I wanted you to know that I invited Vicki Pointer to have Christmas Eve dinner with us. She is alone since her children couldn't come home, so Harry and I thought she should be with us. He doesn't get to see his niece much now that he doesn't go into the store."

"That's great!" I said somewhat distracted.

"She's closing the store early to be with us," Mother added.

"I really must go, Mother," I said as I watched everyone else working. "I'm having the Water Wheel Restaurant deliver some food before we close so we can have our little employee Christmas party here at the shop. Remember before Sam died, we had them over at the house, and Sam cooked a delicious dinner for them?"

"Yes, I do. You have always considered them like family which is so nice," Mother noted. "They were also invited to your Christmas house party, and now you are having some for your Christmas dinner, right?"

"Yes, because they are my family. Now more than ever," I confessed.

"Okay, I'll let you go, honey. See you tomorrow evening," she concluded. "Have fun!"

Sally was taking one call after another, and Abbey and Jean were working in the design room. Sally hung up and looked so sad.

"Can you take care of this one, Anne," she asked handing me the order slip. The tag for the casket spray was *Our Dear Mother.* "The children's are supposed to match the mother's casket spray. Evidently the mother loved red so I gathered that they want them all to be cheerful." I took a deep breath as I thought of the task at hand.

"Do we have plenty of red in stock, especially roses?" I asked Sally.

"I doubt it, Anne. This funeral is going to be huge," she sadly stated. "I'll call for more roses right away."

Thank goodness we didn't have a Christmas wedding scheduled. Christmas parties were ending, and arrangements and orders were being picked up today. Sally was going to open for a few hours on Christmas Eve morning to make sure we could accommodate everyone.

The very sad occasion created a silence in the shop that was hard to describe. This was usually a jovial day as they knew they would get their bonus and share refreshments later. I went to my computer and sure enough, there were out of town orders coming in. I wanted to put my head on my desk and cry when I read some of the messages to be put on the gift cards.

Two little children who were going to have a marvelous Christmas were now gone! I suppose it was a blessing God took the three of them together. This would change the dynamics of the Forester family forever.

When I joined the others, Sally was in the restroom when I answered a phone call from the Kiddie Korner Nursery School where the four-year-old little boy had attended school. The woman was weeping as she tried to tell me what she wanted to order. She said that he had loved Batman, so if I could do anything to create something like that for him, the family would love it. I assured her that we would take care of it, and when I hung up, there were tears flowing down from my own eyes. David watched as I tried to hide my emotions.

"It's the pits, Anne," David said trying to console me. "I saw some of the family come in to Barristers when I was making the delivery this morning, and it was pitiful! Did you know any of them, Anne?"

"No, and I can't imagine how I would cope if I did," I said blowing my nose. "It's Christmas time for heaven's sake! I don't think anything has ever affected me like this especially with not knowing the family!"

I now saw mist in David's eyes as well.

"Keep in mind, Anne, that when we send these flowers from family and friends, it will make them smile," he consoled. I looked up at this gentle, caring guy.

"Thanks David. You're right about that," I said forcing a smile.

"You are so well suited for this business, Anne," he continued. "You have flowers with a heart."

"Back to work," I told myself. What a sweet thing David just said.

As the day continued, our day was interrupted with tokens of joy. Nick brought over a tray full of cookies and fudge, and Mrs. Lane brought us some of her home-made divinity. Aunt Julia always made lots of what she called nuts and bolts each year, so she brought us a big bowl over to enjoy later. The salty mix of cereal, pretzels, and nuts was a family tradition. I encouraged her to join us when she closed the shop.

The news of the accident had spread around town quickly. I took a call from Sue at the funeral home which I knew would relate to the accident.

"I'm calling for Richard and Nancy, Anne," she explained. "They want me to order a mixed arrangement to be delivered for the Forester family. They evidently knew them quite well. It's just awful around here today."

"Do you know how the father is doing, Sue?" I asked with concern.

"Not well. I don't think he is conscious," she reported. *How awful*, I thought.

"I'm sorry we won't see you tomorrow at Mother's," I remembered. "Enjoy your dinner at home for a change."

"I will try, I just hope I don't mess up anything," she cried.

"Give Eli and Mia a hug for me," I instructed her. "I have some things under the tree for them."

"I will, and please tell my part-time Brown's Botanical family Merry Christmas, will you?"

"Merry Christmas to you all as well!!"

# CHAPTER 49

At five o'clock the Water Wheel delivered our roast beef sandwiches with all the trimmings. They placed a lovely holiday spread on top of our design table which made it quite enticing for us. Abbey locked the door behind them when they left, and David started pouring wine for everyone.

"I can only stay a bit as I'd like to pop on home before the nutter arrives," Jean said to me.

"Why would he care if you were late?" Sally asked sharply.

"The holidays can make a man like Al take a turn," she answered looking away. We knew what she meant, and Sally regretted asking.

It was good to see her join us with a glass of wine and some food.

"Before you all indulge yourself or leave, I just want to tell you again how much I love and appreciate each one of you. Since I know money talks around here, you'll find that

Santa has left something in your stocking. David, you will find yours in Kevin's stocking. I will take care of Kevin later." They all laughed.

"Anne, we all have a little something for you as well," chimed in Sally. "You know you are the hardest person in the world to buy for, but with our big hearts and little pocketbooks, we wanted to contribute in some way for you to have that getaway you keep mentioning. We purchased a voucher at the Colebridge Travel Agency that will help you get to where you're going!" I was stunned.

"How sweet," I said, feeling speechless. "You are always trying to get rid of me! Thank you very much! I promise you it will be somewhere!" Everyone clapped and starting filling their plates. For a wonderful block of time we were all able to forget the very sad day we had all just experienced.

Jean was the first to leave as she generously thanked me for everything. Kevin then showed up to join us, and we told him that someone robbed his Christmas stocking. It wasn't long before he and Abbey went on their way. David then mentioned that his wife's family was in for the holidays, and that he would have to leave as well. Sally took her time leaving and expressed her concern about what Jean might be dealing with at home.

"You're so lucky to go home to Sammy," Sally said with nostalgia. "It's times like this that I feel like such a failure when it comes to my personal life. Tim was the closest person I ever felt like sharing my life with, and he took off!"

"Oh, Sally, don't do this to yourself!" I said in defense. "It's the wine talking now, plus you are one of the smartest and most creative people that I know. The right person just hasn't come along! I won't disagree that the holidays can be the pits. Last year at this time, I was in mourning for Sam. I didn't think I could carry on in life, but human nature provides a way. I'm sure the Forester family thought they had it all until a couple of days ago." She nodded, and then we hugged good-bye.

As I cleaned up the place a bit, I was reminded again how blessed I was with this shop family and my lovely son. I had expected the party to last longer so I figured I had time to drop off Christmas gifts for Nancy's children, Amy and Andrew. I called her cell to tell her that I was coming, and she told me she was at the funeral home helping Richard.

"I don't want to bother you then, Nancy, if you all are busy," I added. "I just wanted them to have their gifts along with the other things they open on Christmas morning."

"Very sweet of you, Anne," she acknowledged. "Just drop by my office instead of the house if you don't mind. We are all needed around here with the holidays and this Forester funeral. I have a gift in the car for your little family as well, so this will work."

I agreed and told myself to just run in and out very quickly. I locked up the shop and noticed that it was going to be a busy night for Christmas shoppers on Main Street.

I did find Nancy in her office. She was on the phone and motioned for me to sit down. She looked very stressed from her long day. I put the presents down on her desk as she hung up.

"This Forester funeral has been horrible to deal with, Anne," she stated almost in tears. "We know the family so Richard is at the hospital right now with the father. They don't think he'll make it." She wanted to totally break down but held herself back.

"I don' want to keep you," I said, feeling helpless about the whole thing.

"We have another funeral going on as well so we are really maxed out with help," she explained. "Richard's father even came in to help with a few things. Oh, Anne, when I think of those little guys in that accident, I can hardly stand it!" She took a deep breath. "Here are your gifts, my dear friend. I hope that you and your family have a wonderful dinner. Mrs. Barrister is cooking so at least I don't have to think of all that."

"We're ending up with a pretty big group. I'm in over my head as always," I admitted. "I still haven't heard from Helen. My guess is they're not coming."

Just then Nancy was paged to go to the conference room so I jumped out of my seat to leave as well.

"Sorry, Anne. Thank you for the gifts," she said, giving me a kiss on the cheek. "Let's catch up after Christmas, okay?" I nodded with teary eyes.

# CHAPTER 50

Itook Nancy's bag of gifts and walked toward the parking lot entrance. It was a total surprise to see Jack Fletcher coming in the door.

"Anne, is everything okay?" he quickly asked.

"Oh sure!" I responded being shocked to see him. "I just dropped off the Barrister's Christmas gifts since I wouldn't be seeing them." He nodded. "What are you doing here?"

"As it turns out, the Forester family didn't have a church affiliation so Nancy asked if I would come here to talk to some of the family. I assume you know about this horrible tragedy. There are many friends and family to deal with."

"I do know," I said sadly. "When all the orders started coming in, we were overwhelmed and very saddened by it all." I started tearing up. "We didn't know where to start."

"Let's sit a minute over here," Jack suggested as he looked at the couch along the corridor. I followed.

"We've never had to deal with a whole family's funeral all at once like this," I admitted with my hands on my forehead. "It sounds like Mr. Forester may not make it either which would really be tragic."

"I just came from the hospital to pray with him," Jack revealed with a sigh of hopelessness.

"What a horrible Christmas for that family," I added in tears. "They will never be cheerful at Christmas again."

"God's time table is not ours, and when he calls us there is a reason and a message to be learned," he said softly as he put his hand on mine. Was he giving me a personal sermon? "The Barrister's, Brown's Botanical, and spiritual leaders are special people with gifts to get them through this. God is good, no matter how tragic life can be." I thought of Sam, and tears were now rolling down my cheeks.

"I don't even know the Forester's, Jack," I said as I pulled my hand away. "I know it's hard to loose someone you love, but this was a whole family."

"We are all God's children, and he made us to care for one another," he consoled. "Your concern over this accident says a lot about you, Miss Dickson. We will all pray for their comfort, as I'm sure many did when you lost your husband." Now I did break down into an official cry. Where did this come from? "I'm sorry, Jack."

He got up, went to a water cooler nearby, poured me a cup of water, and returned with it to me. I took a sip trying to get a grip on my emotions.

"I really need to let you go, Jack," I said with embarrassment. "How does one extend holiday greetings at a time like this?" He smiled.

"Christ's birthday will be celebrated each year no matter what may be going on in this world," he said with comfort in his voice. "We have to keep things in perspective and know that there is something to always be joyful about, like one's faith in Jesus Christ." I stared at him.

"I am totally comfortable in wishing you and your family a very blessed Christmas," he said with a big smile.

"Thanks, Jack. I needed to hear this today," I said softly in almost a whisper. "I don't think it was an accident my running into you today." He smiled.

"My good fortune as well!" he responded.

"So, do you have plans for Christmas dinner?" I asked without thinking. The question didn't come out right.

"I'm invited to the Halfords's residence after Christmas morning services, but then I'm on my own for dinner which is perfectly fine with me." I took a deep breath knowing I had to finish my spontaneous request.

"Would you like to join my family for Christmas dinner?" I asked wiping away my tears. He looked at me like he wasn't sure he had heard me correctly.

"Do I sense some kind of sympathy here?" he asked with a half laugh.

"Well, you have to understand that my dinner guests are all quite diverse, and you would probably fit right in!" I said breaking into a big smile.

"Are you making fun of me, Anne Dickson?" he asked standing back to give me a flirtatious smile.

"I'm sorry. I'm just an emotional basket case today," I admitted feeling stupid. "I have put you on the spot and taken advantage of your time and sympathy."

"I'd love to. What time?" he said now looking serious. Oh dear, did he just accept?

"Oh, about six o'clock would be fine," I quickly responded.

"That works for me," he said nodding. "Are you going to be okay?"

"Of course. Please be on your way. I must get on home," I said getting up to leave. He nodded and gave me a pleasing smile.

He walked away leaving me in disbelief as to what had just occurred. Thank goodness Nancy didn't see us talking.

On the way home, I pictured my dinner guests and what they might think about my inviting Reverend Fletcher to join us. Of course my Mother would be pleased, or would she? Hmmm.

# CHAPTER 51

Ella and Sammy were playing in the den and happy to see me. I brought leftover goodies from our shop party, which Ella refrigerated while I played with Sammy. He was in a cheerful mood so we had to go through his favorite playtime rituals which made his laugh come deeply from his belly.

When Ella returned, I told her about the accident. She, too, did not know of the family, but it struck her deeply with sadness. I then told her about going to the funeral home to drop off gifts which lead to my conversation with Jack. I presented the dinner request as if I felt sorry for Jack, and that he had surprisingly accepted.

"Goodness! What brought that on?" she asked sitting down. "I didn't realize you had gotten to know him that well."

"I really don't know him, but I was in such an emotional state after talking to Nancy. We both sat down for a chat, and out of curiosity I guess, I just asked him if he had any dinner plans. I could hardly believe I did it myself! He was being so nice, and I guess I felt sorry for him."

"Come to think of it now, didn't you dance with him the night of your party?" Ella said with a hidden tone.

"He cut in. What was I supposed to do?" I defended myself. "Ella, he is an interesting man, I do have to say. I do think he is indicating some interest in me, but he's not pushy in the least."

"I'm not surprised," she answered right back. "You're a pretty interesting person yourself!"

"I don't think I'm ready for that, Ella," I said with a pang in my stomach. "Sam still has my attention every hour of every day. How could I possibly want to change that?"

"I think you need to relax and just be his friend," Ella suggested. "Goodness knows he is new to the area and he's probably surrounded with more senior citizens at that church of yours than folks your age."

"Well, Nancy and Richard have certainly taken him everywhere you can imagine so he's had plenty of chances to meet other women," I declared. "I did see him have lunch with some woman last week."

"Well, beat yourself up, honey, but now you are the one that has him coming to Christmas dinner!" she teased. "I say you certainly know how to make everyone feel welcome! God bless you!"

"Do I tell Mother ahead of time or not?" I asked feeling confused.

"I don't know why you should!" Ella responded. "She may make more out of you doing that than by him just being here."

"You're right!" I nodded. "Ella, what would I do without you?"

I went upstairs and started Sammy's bath with thoughts running wild about my awkward invitation. After his bath, he looked pretty tired, just like his Mother. I decided there would be no story tonight, and he didn't give me a bit of fuss. I kissed him goodnight and went to my bedroom.

Suddenly, I picked up the aroma of fresh lilies. I knew that smell very well. Sure enough, on my dresser was a lovely bouquet of lilies, most likely from Grandmother's spirit. I hadn't seen this acknowledgement for some time. What was this about? It was then I noticed something strange. The vase was next to a wonderful framed photo I had of Sam except the picture was face down. How did this happen? I immediately picked it up to sit it upright again. Surely Grandmother wouldn't do anything like that? Perhaps it fell forward when Ella dusted the dresser.

I dressed for bed feeling emotionally drained. I was about to turn off my bedside lamp when I noticed that the wedding photo of Sam and I was placed face down as well. I quickly set it upright. It was a strange occurrence which now seemed to have been done on purpose.

"If this is one of your pranks, Grandmother, it is not a funny one," I said out loud to her. "I will always love Sam, and these photos are a comfort to me."

It was hard for me to fall asleep despite my exhaustion. Was Grandmother telling me to put Sam out of my mind? Surely not! Aunt Julia always claimed she was mean, and I defended her, of course. What really was her message?

# CHAPTER 52

Waking up the morning of Christmas Eve day was exciting; after all it was my son's first Christmas. I even had pangs of excitement as I reflected upon my childhood this very day. Tonight we would make that story book trip to Grandmother's house that we read about. There was quietness in the house which usually meant that Sammy was still asleep. I had a lot to do after breakfast so I needed to get an early start.

Picking up the chocolate cake from Nick had to be first on my list, and then I had to check in on Sally who was tackling the day alone as she worked on the Forester orders. Christmas Eve was typically slow, but this year was different. I once again thought of the Forester family waking up to what was supposed to be a wonderful Christmas. I hoped God's grace took their lives quickly without pain. I had to stop thinking about this, or it would ruin my Christmas as well.

I put on jeans and a heavy sweater before brushing my teeth. When I went by my dresser to grab my purse, I noticed that Sam's picture was once again placed face own. This made me angry, and I didn't have time for a foolish response to a crazy ghost!

I checked in on Sammy and realized that Ella must have taken him downstairs earlier. When I got to the kitchen, Sammy was picking up cereal from his tray. He started laughing as he once again threw his sippy cup on the floor to get my attention. I tried not to react because it was what he was expecting.

"No, son," I warned putting the cup away from him. "Don't you know Santa is watching to see who is naughty and nice?" I kissed him on the cheek as he gave me a puzzled look.

"I have a nice plate of Sammy's hand-print cookies ready to hand out," noted Ella. "You'll take a couple tonight of course. They will love them. I have their gifts by the front door so don't forget them."

"Good idea," I stated. "I will do the same."

"When Sammy's down for his nap this afternoon, I'll set the table," Ella revealed. "Did you see the size turkey and ham we have? It's enough to feed an army. I have the dressing and the vegetable casseroles in the refrigerator. We'll have the mashed potatoes and sweet potatoes to deal with at the last minute of course. Aunt Julia insisted on bringing her cranberry Jell-O salad. If I remember, it's quite tasty."

"I don't know when you did all this, Ella," I bragged. "Yes, I love her cranberry salad."

I shared a slice of English muffin that Ella had already toasted and poured myself some coffee. My cell phone was ringing and I saw that it was Pat. I panicked at first thinking something may have happened to Helen.

"Good morning and a Merry Christmas," Pat cheerfully greeted me.

"And the same to you!" I manage to respond.

"If you are still up to your invitation, do you suppose you could arrange for someone to pick up Mother and me at the airport tomorrow around noon?" Pat asked. I couldn't believe it!

"Why sure!" I said right away. "I can't believe you're going to make it after all! I had pretty much given up on you two."

"We wanted to wait until the last minute to see how she felt," Pat explained. "She decided she didn't want to miss that first Christmas with her grandson, so we're coming!"

"Wonderful," I responded with excitement. "I'll give Uncle Jim a call right away to make the arrangements. If he isn't available, I'll have Kip pick you up. The weather here is holding, so no snow or ice! I can't wait to see you both!"

"Are you still going to your Mother's house tonight?" Pat asked with interest.

"Oh yes. There are traditions that have to continue around here!" I assured her.

"I can't wait to see everyone," Pat added. "Please don't make a fuss with us coming on such late notice."

"I planned for this surprise, I'll have you know," I said before hanging up. This was very good news!

Ella heard the conversation, and I could tell her mind was going wild thinking about what to do next.

"It's a good thing we went ahead and counted them in on the table setting," she said with relief. "Their room is ready too, thanks to you, Anne."

"I think with so many sitting at this big table, I am going ahead with place cards," I said heading into the dining room. "Oh, I mustn't forget the beautiful centerpiece Abbey did for the table. I better get going, Ella!" I headed towards Sammy to kiss him good-bye.

"Be on your way. We'll be fine," she always said. "Just don't forget that dessert of all things!"

While in my driveway, I called Uncle Jim to see if he could pick up Sam's family. He told me that he would be at Rose's house. I gave him credit for not asking if he could bring her to my dinner. My next call was to Kip, who said he was more than happy to do so. I gave him the flight information and off I went.

When I arrived at the Bakery, Nick was thrilled to see me and had included a plate of his small cannolis which he knew I loved. He could be such a generous guy when he wasn't complaining. I gave him a hug, which embarrassed him. I told him he was a diamond in the rough.

When I got to the shop, Sally was arranging a piece for the funeral as I suspected would start happening. I helped her arrange them in the walk-in cooler for easy access the day after Christmas for the funeral.

"I haven't seen anyone here in the shop but Kathy. She came in for a poinsettia to take as a house gift," she reported. "Don't forget your centerpiece for tomorrow, Anne."

"Thanks, Sally," I said observing that she looked a little sad. "It's a good thing I have folks looking out for me! So what are you doing tonight?" She wouldn't look up.

"Not much, but I am looking forward to your delicious Christmas dinner!" she reported more cheerfully. "I am actually invited to a small gathering this evening, but don't think I'll go."

"Helen and Pat are coming, by the way," I noted. "They just called."

"Well, that's a nice surprise," she said smiling as she continued to work.

"We have another surprise for the dinner as well!" I announced.

"Oh, and what would that be?" Sally asked patiently.

"In a weak moment at the funeral home, I invited Jack Fletcher," I said taking a deep breath.

"The Reverend Jack Fletcher?" she asked sarcastically. I nodded. "Well, he better not pay a fancy to you at the dinner like he did at your party because your mother-in-law will pick up on it in a flash," Sally warned.

"Oh, I really didn't think of that at the time," I said with frustration. "Holy cow! Well, maybe she'll be impressed that Mother and I felt sorry for him and extended the invitation!"

"Right!" Sally said grinning. "We'll see about that!"

# CHAPTER 53

The frigid temperatures were nearly at zero, but the good news was there wasn't any precipitation. Ella, Sammy, and I dressed as festively as we could while staying warm and comfortable. I was relieved that Mother and Harry didn't have to go out in this weather.

Ella made a Christmas stollen to take for their Christmas morning breakfast. She kept one for our little family at 333 Lincoln as well. I remembered how much I had enjoyed it last Christmas. Ella took each of the Sammy hand-print cookies put them in individual plastic bags, and tied them with red ribbon. I knew they would be a great hit wherever they went.

As we approached my old neighborhood, I noticed that it looked as festive as always, with the same neighbors decorating it the way they had done year after year. Even Mr. Carter managed to have lights on his big pine tree despite the passing of Mrs. Carter. I suggested to Ella that we leave a cookie for Mr. Carter when I realized my neglect to get him some kind of gift. She assured me that we had plenty to share.

We were greeted with open arms as Mother and Harry opened the door. Sammy was immediately taken from us and was quite intrigued with all the fuss.

"Well, we know who rates here!" I told Ella.

Harry took our coats and we headed to the nice warm fire in the living room. As I warmed my hands, I wished for the real fireplace that we once had when I was growing up. I would always have to help carry in the firewood and loved to sit and watch the logs crumble as they burned. After my father died, Mother replaced it with a gas stove.

"Vicki, how nice to see you!" I greeted Vicki Pointer from Pointer's Book Store. "How have you been?"

"Very well, thanks," she happily responded. "I can't believe how much Sammy has grown! The last time I saw him, he was just a few months old."

"It's his first Christmas and it may as well be my first, it's so much different," I claimed.

"You've had quite the year, Anne Dickson," Vicki stated with a big smile. "Your expansion is beautiful at the shop and some day, I would like a tour up on your hill."

"Anytime," I said now accepting a cranberry drink from Harry.

"My own concoction, Anne," Harry bragged. I took a sip.

"Delicious!" I said taking another sip.

Aunt Julia, Uncle Jim, and Sarah all arrived together which surprised me. I wondered what his Miss Rose thought of all this? As soon as they took off their coats, they headed straight towards Sammy. Ella kept a close watch, and

I noticed that Sammy didn't take his eyes off of her as many begged for his attention. I stopped to take advantage of this Kodak moment by taking some photos on my cell phone.

"That was a wonderful Christmas party!" Uncle Jim said as he walked near me.

"It was really fun," I agreed with a smile. "It's a tradition I would like to keep going. I look forward to a nice dinner tomorrow night, as well!" He nodded.

"I always appreciate being included, as you well know, Anne," Uncle Jim reminded me. "Rose loved your house by the way. She is still talking about what a great evening she had."

"So, is Rose a serious venture, or better said, relationship?" I asked boldly.

"Not at all," he quickly answered. "She is quite fun and I've discovered that I'm not very good alone like some people are. I try to help Sarah and Julia at the shop whenever I'm needed, and I occasionally get fed." We both chuckled. "By the way, I am happy to help you any time as well, Anne."

"I know, but I am blessed with a good staff that goes well beyond their shop duties," I explained.

"Sammy is going to need some male influence in his life, and I hope I come to mind when you feel the situation is right," he advised. I nodded with some confusion. Did he mean when it was time to throw a baseball or what? Was Sammy in an unbalanced life of all females?

Harry was now getting our attention and told us to gather around the table. Sammy was put in my old high chair that I used as a child. Harry clinked on his glass to quiet us down.

"Thank you all for sharing this Christmas Eve dinner with Sylvia and me," Harry began. "Having you here with us, Vicki, is especially meaningful to me. As you can see, I became lovingly connected to this wonderful family when I married this dear lady." He looked right into Mother's eyes and she blushed. "Please join me in prayer." We joined each other's hands. I took Aunt Julia's and Sammy's hands. "Thank you for this food and the many blessings you have bestowed upon this family. Tomorrow we will celebrate your birthday with joy and goodness of heart. In your name we pray, Amen." We all responded by saying amen.

Mother insisted that we all sit down while she went in the kitchen and came out with a beautifully sliced pork roast, just waiting to fill our plates. She then brought out bowls and bowls of goodies in our family style tradition.

Sammy loved mashed potatoes, as did I. He surprisingly ate a couple of bites of sweet potatoes which was unusual. I didn't dare let him get hold of his sippy cup, or it could land just about anywhere on the table or floor. I don't know when he would start learning table manners. I wasn't sure, but I knew my Mother was a stickler for them.

Mother was using Grandmother's set of Havilland china as well as her crystal. She was very picky about when she would use any of Grandmother's things. The silverware was a lovely shell pattern. Mother said they had received it as a wedding present. My floral centerpiece was perfect on her off- white lace tablecloth. Would this be the last time Mother would be entertaining like this? I hated to think about it.

The conversation around the table was pleasant. It ranged from how well Harry was doing with his health to the latest news from Pointer's Book Store. I knew my luck with keeping Sammy busy at the table was running out, so when I picked him up out of the chair, Mother announced that we would have coffee and dessert in the living room. Harry announced that he would play Santa and give out the Christmas gifts as he proudly put on a Santa hat. Everyone got a kick out of it and we all followed him to the living room. I set Sammy on the floor where almost anything could come his way and be entertaining. I couldn't think about eating a dessert with all the wrapping paper and attention coming Sammy's way. Uncle Jim was thoughtful enough to get some of the action on video on his phone. The joy I was experiencing would be ten times nicer with Sammy's father watching. Oh, how I hoped he was!

# CHAPTER 54

Toy after toy was coming Sammy's way so I joined him on the floor to help him tear off the paper. Uncle Jim was catching great videos on his phone. Everyone knew about Sammy's love for trains; Aunt Julia and Sarah made sure he received one. It was now his favorite. Opening gifts used to be so much more orderly, but tonight's activity seemed to be out of order, and I knew my son had a lot to do with it.

Mother gave her gift bag directly to me which I knew had to be the quilt she was working on for Sammy. I stood up and pulled out the intricately appliqued train quilt which got everyone's attention. They all clapped and showered Mother with compliments.

"Sammy, look what Gramma made for you!" I bragged. He looked for a second and then went back to putting part of his new plastic train in his mouth. "Mother, it's just wonderful. What a treasure!"

"I'll have you know, Sarah and I put a few stitches in there, too, and even Sue!" yelled Aunt Julia from across the room.

"That was a fun quilt to do!" responded Sarah.

"I don't know how any of you do all that, but the quilt is amazing," voiced Vicki.

"They are all good quilters now, Vicki. thanks to Aunt Julia's first quilt when we all learned together," I informed everyone. "Thank you all so much! I love it and so will Sammy!"

"My wife amazes me with her talents every day!" chimed in Harry.

Now Uncle Jim was arranging everyone for a photo shoot that included the quilt. I loved the one I took with my camera of Sammy, Mother, and the quilt. Good Kodak moments to remember as Mother always said. I then noticed she had sewn a label on the back corner of the quilt that said, "Sammy's First Christmas, the date, made with love, from Grandmother." I gave her a hug knowing how much time and love she had put into it.

When we gave Mother and Harry Sammy's cookie ornament, Mother nearly cried. I had to pull out my phone and show her a photo of what Sammy looked like making the cookies, and they fell apart with laughter. She went right toward the tree and placed it right in the center for all to see.

"Any gifts from Grandmother?" Aunt Julia asked with a tease. I could not tell her about Grandmother's latest prank with Sam's photo, but I wanted to.

"Not yet anyway," I answered.

"Anne, here's a little something for you," Mother said presenting me with another gift bag decorated in red and white. "I hope you like it." I pulled out a beautiful black cashmere sweater with pearl buttons. It was one of the most gorgeous things I had seen in a long time. I was truly surprised and knew Mother had to pay a good deal of money for it. I held it up in admiration.

"I don't know what to say, this is awesome!" I said as I ran my fingers over it with my hands.

"Can I feel it?" asked Sarah.

"I'm so glad you like it," Mother said beaming. "We can return it for something else, but I know how much you love your cream sweater."

"Love you, Mommy dearest," I said giving her a hug.

My gift cards to everyone did not cause excitement, but everyone seemed to be pleased. To my surprise it was getting past Sammy's bedtime and he was still having a ball showing no signs of crankiness. I had Sarah to thank for most of Sammy's attention as she demonstrated all his new toys for him.

"I want to tell you, Anne, how pleased I am that my Uncle Harry married your mother," said Vicki out of the blue. "I was so worried about him, but I feel so much better now that I see him so happy and cared for. I hate that she is probably tied down a lot."

"She wouldn't have it any other way, Vicki. Aunt Julia and I are trying to make sure she gets out and about. How is the book store doing?"

"Really well," she reported. "How's that Taylor House book coming along?"

"It's not, I'm afraid," I said with a sigh. "I've not written much since Sammy arrived. I feel like I'm living the book right now." She laughed and nodded with approval.

There was a loud, sudden knock at the front door like an urgent call of some kind. Harry opened the door immediately, and to our surprise, voices started singing carols on the front porch. Aunt Julia then told me that she had arranged it with some of the carolers from the street. I picked up Sammy and wrapped him in his train quilt as we stood by the door to join them in singing Joy to the World. Happy faces were everywhere as Harry hugged Mother for warmth, and Uncle Jim even put his arm around Aunt Julia. It was so emotional and a very clever idea coming from Aunt Julia. Ella and Sarah joined in singing as they made funny faces at Sammy. It was a magical, unexpected surprise on Christmas Eve. Before they went away, Uncle Jim gave them a generous tip and waved them on with the song, We Wish You a Merry Christmas! It was the grand finale to a great Christmas Eve, and we all began to head home.

# CHAPTER 55

Christmas morning I wanted to turn over to where Sam used to lay and wish him a Merry Christmas. That sick feeling once again hit me with reality as I faced different circumstances this Christmas. I think it was natural to look back on one's childhood on a holiday like this. The memories are vivid, and they are joyous for most people. Like other children, I would wake up early and want to jump out of bed at the least indication of morning to see what Santa had brought. Perhaps in the next year or two, it would be Sammy doing the very same thing. I turned over to relish the thought and perhaps sleep a bit longer since I knew it would be a very long day. Then I noticed that my wedding photo was turned down again. I sat up to rearrange it feeling frustrated. *Ignore her*, I said to myself!

I looked at the very early hour on my clock and decided to get up. Today would place a heavy load on Ella, and I needed to be helpful. When I remembered the afternoon arrival of Helen and Pat, I had mixed feelings of sadness and joy. Seeing Sam's mother brought so many memories. I must remember she has increased pain in losing a son, versus, that of losing a husband like I had. We would both find this holiday hard to get through.

I looked outside the window in case we might be surprised with a white Christmas, but in the darkness, there just appeared to be the bitter cold of winter. With a greater chill in the house, I put on my heavy, white terry cloth robe, brushed my teeth, and headed toward Sammy's room.

He was sound asleep. He was undoubtedly exhausted from his first big Christmas Eve at his grandmother's house. I leaned over and whispered Merry Christmas! As soon as I did, there was that smell again of Sam that told me he was also there with me. It gave me a sense of warmth and love for some strange reason. Without thinking, I turned thinking I might get a glimpse of him. Wishful thinking did not make that happen.

I quietly walked down the stairs looking at the beautifully lit tree that had special meaning today. Surprisingly, Ella was not up so I started the coffee. I would wait and join Ella with some of her tempting Christmas stollen. We had decided that we would have our little Christmas with just the three of us before the big festive feast preparations began. I wondered what the rest of my family and friends were doing about now. Harry and Mother were likely sleeping in from their big night. Mother had mentioned they were going to the late church service and asked if I would be there. Could I possibly fit that in this morning? Would Jack notice if I wasn't there? What would he think if I wasn't? Did I care what he really thought anyway? Was Sue sharing Christmas morning with just her two little ones, or was Devin now included? I wondered if Aunt Julia and Sarah were up yet. Was Uncle Jim waking up with Miss Rose? Hmmm.

"Merry Christmas!" Ella said joining me in her robe.

"Merry Christmas to you as well, Ella!" I said as I cheerfully greeted her. "Is Sammy still asleep?"

"He sure is!" she reported. "He had quite a night, didn't he?" I laughed and nodded. "He kept his spirits pretty good since our visit was way past his bedtime."

"I'm thinking about going to the late Christmas service Ella," I suggested. "Do you think it's possible?"

"I thought you might!" she said putting the big turkey into the oven. "I think that new reverend will be looking for you!" She gave me a wink. "We are in great shape until Kip delivers the Dicksons this afternoon. Things might get a little hectic then."

"Please don't read too much into this reverend business, Ella," I reminded her.

Just then we heard Sammy yelling out, so I went upstairs to change him into warm flannel pajamas with feet to keep him warm downstairs.

When I came downstairs, Ella had us set up for a fancy breakfast near the fire in the den. She brought in Sammy's chair so he could join us for breakfast.

"This is special. Ella," I complimented her. "I haven't used this tea set for some time."

"It looked festive with the green holly, so enjoy the stollen and your coffee before we open a few gifts," Ella suggested. We even shared bits of the stollen with Sammy.

"Here are a couple of things from me, Anne," Ella said handing me the gifts. "This one is for Sammy, but he may not think too much of it. He has more toys than he knows what to do with so my gift isn't very exciting." I pulled out a white cardigan sweater that Ella had undoubtedly knitted for him.

"Oh, Ella, this is so darling," I said holding it up. "You do such lovely work."

"This one is for you and I'm keeping my fingers crossed that you will like it," Ella added.

I opened the next gift and knew it was going to be something she made for me. It was a gorgeous white cardigan sweater out of the same yarn as Sammy's sweater.

"Wow, Ella," I exclaimed. "When did you have time to knit this?" I stood holding it up to me.

"Well, you know I watch a lot of TV, and I just can't sit there and do nothing!" she explained.

"Thank you, thank you, Ella," I said, giving her a hug. "We love you! Now here's something for you!"

She first opened the gift card for the knitting shop she frequented. I could tell she was thrilled. Inside was a check that totally surprised her.

"Anne, this is way too generous," she said about to cry.

"You deserve it, Ella," I bragged. "I don't know what I'd do without you. I hope you take a trip or buy something extravagant!" She laughed. "Now here's a little something from Sammy boy!"

"More?" she gasped.

She opened a small box that contained a silver watch. She looked at me in disbelief.

"It's beautiful, but I have a watch, Anne. I can't accept this!" she claimed.

"Didn't you say you'd had that since high school?" I teased. She laughed and nodded. "Well, now you can put the old one in the archives or hand it down to someone special."

"Better not say that or I might give it to you, Anne," she teased back.

She was now wiping tears from her cheek as she gave Sammy and me both a big hug of thanks.

# CHAPTER 56

I'm afraid Santa was frugal with Sammy this year," I sadly told Ella. She now sat down with her cup of coffee as if I had an explanation. "I did write a little poem the other night thinking about his first Christmas. It's kind of silly, but I thought in later years he would get a kick out of it. I'll write it in his baby book."

"You did?" she questioned. "I'd like to hear it, or is it pretty private?" I smiled and shook my head as I pulled the piece of paper from my robe.

"Here goes!" I began to read.

SAMMY'S FIRST CHRISTMAS

Little does he know, this is Christmas number one.

Little does he know, it's just Mom and son!

Little does he know, a Father's missing here.

Little does he know, a Santa may appear.

Little does he know, it's his first Christmas tree.

Little does he know, how special this can be.
Little does he know, all, is in his baby book.
Little does he know, before he has a look!
Little does he know, how happy he will be.
Little does he know, how much joy he brings to me!
Merry Christmas Sammy!

"Oh my, Anne, how special that is!" she finally said. "You wrote this from the heart, and in time, this will be so meaningful to him. I think it's a precious gift. Sammy will not only be written about in his baby book, but he is now making history with the Taylor/Dickson house you are writing about. Another generation is having his first Christmas just like the Taylor's little girl. What was her name?"

"Miranda," I recalled. "I'm sure she had her first Christmas here. The Taylors had some happy years here before Miranda died and Albert started playing around. Grandmother Davis loves to overshadow that with her bitterness over not marrying Albert, but I must respect the fact that Marion spent many years in this house. She didn't deserve the unhappiness she experienced with losing a daughter and an unfaithful husband."

"That's right, Anne," Ella said sadly.

"Oh, Ella, it's getting late," I said looking at the clock on the mantel. "I think I will hurry and try to make the late service. I'll dress Sammy when I return. He has to look adorable for his Grandmother Dickson and his Auntie Pat."

When I went upstairs to dress, the photographs were once again turned face down. If Helen gets a glimpse of this, she will be horrified. I set them back up and told myself Grandmother was not going to ruin my Christmas spirit. So far, the day was off to a pretty good start.

# CHAPTER 57

I barely made it to the church on time. I found Harry and Mother sitting near the back, where there seemed to be room for me, so I joined them. Mother looked delighted when she saw me.

"Merry Christmas you two!" I whispered as I sat down. "Nice hat, Mother!" She smiled. Mother always wore a hat to church since I can remember.

I was pleased when I started reading the church bulletin that traditional Christmas carols were listed, as well as a solo selection sung by their amazing choir.

Seeing Jack in his robe helped remove any previous thoughts and memories of our conversations. Even the dance we shared was a blur. He was handsome, I'll give him that. Hopefully, he would not spot us sitting back here.

After we sang a couple of carols, Jack started telling the Christmas story which we never get tired of hearing year after year. The first time it became significant to me was going to Sunday school in the basement of this church. I supposed my little Sammy would be doing the same someday. Jack would be meeting Sammy for the first time this evening. What would he think?

As we were leaving the very inspiring service, Jack and the Assistant Pastor were at the door to wish everyone a Merry Christmas.

"Did you all have a good Christmas Eve?" Jack asked Mother.

"Yes, it was extra-special with our new grandson, of course," she replied with pride.

"I'm looking forward to meeting him this evening," he graciously said. "Are you sure I can't bring something this evening, Anne?"

"We're fine, but we may put you to work, you never know," I teased. "I can't promise that Sammy will still be up, but we'll see."

"I hope so," he said as he lightly squeezed my hand goodbye.

Mother had a puzzled look on her face, when she realized Jack would be joining us.

I wanted to get home and quickly answered Mother's question about what time Helen and Pat would be arriving. I blew them a kiss and told them to come over anytime.

On my way home, I decided to drive down Main Street. There were very light flecks of snow that sputtered now and then. The street looked so beautiful with no parked cars and all the shops closed. It was picturesque to say the least. When I got near the shop, I saw Sally's car parked in front. That was just like her to spend her day off in her second home.

I knew she was likely knocking herself out for any funeral orders due tomorrow. I kept driving. That was not going to be happening for me on my Christmas day.

When I walked in the door at home, I could smell the turkey cooking. Ella was with Sammy in the kitchen demonstrating some of his new toys.

"How was church?" she asked as she snacked on a sandwich. "How about a bite to eat, Anne? It will be quite a while before dinner. I'd like Sammy to finish his apple sauce, but he is quite distracted with all these toys."

I took over with the apple sauce and told Ella that we may still have a white Christmas. I took Sammy out of his chair, since there was no interest in food, and took him upstairs to change his clothes. As I trucked up the unending stairs, he seemed to get heavier and heavier each time.

I picked out a cute denim overall with a little red shirt. Tonight he would wear his new red velvet jumpsuit fit for a prince. He was so adorable I sometimes found myself staring at him.

As I brought Sammy down the stairs, I noticed that the lights were out on the Christmas tree in the living room. I put Sammy down on the floor and noticed that the tree was unplugged. How did this happen? I glanced into the living room and saw the striking stockings I had purchased in Door County, and noticed there seemed to be something in Sammy's stocking. I jokingly thought that Santa must have come last night. I reached in and pulled out a little Golden Book titled The Three Bears. It was quite worn. I then reached in and pulled out a small, drawstring bag. I looked in the bag, and there were nickels; just nickels. Who would put this stuff in here?

I was about to walk away to check with Ella when I saw the toe bulging from my stocking. I reached in and pulled out another drawstring bag of coins. When I opened it up, I noticed they were English coins of some kind. Someone was playing a joke on me because Ella and Sam's stockings were empty. I called out to Ella to come in the room.

"Did you put this stuff in our stockings?" I asked knowing it couldn't possibly be her.

"What things?" she asked looking puzzled.

When I showed her, she shrugged her shoulders and shook her head in denial.

"No one has been in this house but me," Ella noted. "Do you know of anyone?"

"Grandmother is the only one who plays games around here, and this one is not very funny," I said out loud. "Look at this pathetic book! If this is all Sammy is getting from Santa this year, he's being pretty stingy." Ella had to laugh, but I was not laughing.

"I'll take these things upstairs," I said to Ella. "Are the flowers in the guest room?"

"Everything is perfect," Ella responded. "I even gave it a light dusting this morning to make sure. Let me take this handsome fella with the red shirt into the den. It's about time for his cartoon show."

I took the mystery gifts upstairs and laid them on the hall table in hopes they would disappear. I peeked into my room to make sure the photographs were still in place before Helen's arrival. To my delight they were correctly in place.

# CHAPTER 58

We heard a car drive up and it was Kip delivering our Christmas guests. Light flakes were still coming down but disappearing in the sunlight. I took a deep breath and said a little prayer that this visit would go well, so Sam would be proud of me.

I put on my coat to greet them on the porch. Kip helped them out of the car. They appeared very chipper as we hugged and kissed.

"The lights are wonderful, Anne," Pat complimented me. "Surely you had assistance doing all of this!"

"Oh, I did, and thank you!" I followed. "I love it, too!"

"I'll take their things upstairs, Anne," Kip offered.

"Thanks Kip. They go in the red guest room," I instructed.

Ella was holding Sammy when we came into the house. The crash of attention made Sammy pucker up as if he were under attack. I made excuses for him as I took their coats.

Helen was very taken back when she claimed how much Sammy looked like Sam. Sammy hung on tight to Ella as they sang their praises.

"Welcome! I have some hot beverages ready for you," announced Ella. "Please go in the study and get warm by the fire. It is frightfully cold today!" Ella took their coats and, we all went to find the lit fire.

I took Sammy from Ella and sat on the couch next to Helen so he could get used to her. She finally managed to get him to smile despite his shyness. She was dying to get her hands on him, so at one point, I just set him on her lap and stayed right there with him. It was so heartwarming to see how much love she was showing to him. I hoped Sam was watching from somewhere.

"I'll patiently wait my turn," said Pat. "I'll bribe him at some point with a new toy."

"Good luck with that after last night's shower of toys at Mother's house," I shared. "Oh Pat, wait till you see the darling train quilt Mother made for Sammy's Christmas present."

"Well, I had to quilt him a little something myself," revealed Pat. "He is my only nephew, and I plan to spoil him. I don't think I can compete with your Mother's work however."

"Oh, I can't wait to see it," I said with excitement. "Quilts are so much work and take so much time! How about some of Ella's delicious Christmas stollen? Dinner will not be for some time. I bet you could use a little snack."

"That sounds marvelous, Anne," said Helen. "I used to make some every Christmas myself."

"How have you been feeling, Helen? How is that other daughter of yours, Elaine?" I boldly asked. "I don't really hear from her much," I noted. Pat and Helen exchanged glances.

"We don't hear much either, Anne. I'm a bit tired from the trip, and if you don't mind, I'll lay down for a bit before the dinner party."

"By all means, and I need to put Sammy down for his nap as well," I said as I picked up Sammy from Helen's lap. "I have you both in the same guest room as before. I hope that will suit you."

"Oh my, yes," answered Pat. "It's such a nice big room. We brought you some homemade candy that Mother and I made."

"Oh, how nice," I responded. "I think I do remember some very delicious fudge when I was at Helen's last visit."

After Helen sampled the stollen, she got up and announced that she was ready to go up. I told her I would go up with her to put Sammy down. We had to take the steps very slowly. It made me feel badly that I did not make a place for her in the maid's room downstairs. It was sad to see more deterioration in her.

We were walking toward the guestroom when she spotted the items I had laid on the hall table. She did a double take before she stopped. She picked up the Golden book for a closer look.

"Sammy is fond of the three bears?" she curiously asked. I snickered.

"No, I don't think I ever read it to him," I said casually. "It's a long story and a strange one at that." She kept staring at it.

"This was Sam's favorite story, and I would swear this was his book," she said with certainty. "Where did you get this?"

"I doubt it, Helen," I hopelessly said. "I found it in Sammy's stocking this morning along with a little bag of nickels of all things. I haven't found the culprit, as yet, but I'm sure there's an explanation." She suddenly sat on the chair on one side of the table.

"Are you okay, Helen?" I nervously asked. She looked like she had seen a ghost.

"I think if you'll look halfway through the book, you'll find a page that's been torn in half," she claimed. I looked at her strangely. "I know that book very well from reading it to him so often. I guess you never knew that Sam always collected nickels as a child."

I sat in the other chair. What was she telling me? I silently looked in the Golden book, and there it was-a half torn page. I wanted to faint.

"This can't be, Helen. I don't know what to say!" I cried. I was now tearing up in frustration.

# CHAPTER 59

Pat arrived at the top of the stairs and wondered why we were sitting there. Helen took the Golden book and put it in Pat's hands.

"Oh, my little brother's favorite!" she immediately said.

"I found it in Sammy's stocking this morning," I revealed. "I have never seen it before. There was also a bag of nickels."

"So no one knows how they got there. Is that what the two of you are saying?" Pat surmised. Helen and I both nodded.

"I also found something in my stocking that makes no sense at all," I shared. "I got a little bag of English coins." I shrugged. "The other two stockings were empty."

"What could that be about?" Helen asked. "How very strange for you, Anne!"

"Stranger things have happened in this house, I can tell you that!" I confessed.

"Weren't you and Sam supposed to go to England before our Dad became ill?" Pat cautiously asked. I nodded.

"The last Christmas we had together, he even gave me red luggage trying to encourage me to travel more," I added. Sammy was getting fussy, and I needed to get him to his room.

"Sounds like Sam would like to send you to England to spend those coins!" Pat stated as if she really knew.

"I must lie down, girls," Helen said with confusion as she went into her room. Pat shrugged and then followed her mother.

This conversation was crazy so I dismissed it. I would have to think about this later. I got Sammy changed and settled for a nap. Needing to think, I went into my room and shut the door. I collapsed on my bed in pure confusion and sadness. Was Sam trying to communicate with me? Was I imagining his smell before? Was he going to be like Grandmother and make objects appear and disappear? How am I going to get through the evening with this nonsense?

I heard a knock at the door. I got up to answer it and saw that it was Pat.

"May I come in?" she asked. "Were you lying down?" She was holding a large bag in her hand.

"No, I just put Sammy down and I was trying to decide what to wear tonight," I said telling a lie. "Is Helen okay?"

"That's what I wanted to talk to you about if you have a minute," Pat explained. "It's truly a miracle I got her here. If it weren't for that precious grandson, she could not have made it. She has many health issues that are very serious, Anne. She's refusing much treatment so this is likely the last visit you will have with her." I felt ill. How was this all going to be a Merry Christmas?

"What is it?" I begged. "What is the problem really?"

"She has made me promised not to discuss one detail with you, Anne, and I must honor that," Pat claimed. "She is adamant about folks not knowing her health situation. After Sam died, she gave up on a lot of things. Know that if she has to sit or lay down, she truly needs to, so I hope you understand. Seeing Sam's little book really threw her for a loop."

"I can imagine as it did me as well!" I said wanting to cry. "Oh, Pat, this has to be so hard on you! Is Elaine helpful at all?"

"That's another subject to avoid, I'm afraid," she said looking down. "We rarely hear from Elaine these days and it hurts Mother so much. I think between Father's death and Sam's, she just doesn't feel a connection with us anymore."

"How sad," I said. "Life is so short."

"Well on a happier note, I wanted to give you this quilt without anyone else around," she said perking up. "I think you'll find it interesting considering what just took place today. Mother found this cute kit and asked if I would make it for Sammy."

I opened the gift bag and pulled out a darling over-sized crib quilt that had The Three Bears cross-stitched on it with an appliqued Goldilocks. It was adorable.

"More of The Three Bears!" I said with amazement. "How did I not know this about Sam? Come to think of it, I don't think I shared that I loved Cinderella growing up." Pat laughed. "It's so beautiful and must have taken you forever to do this! I'm glad it's bigger than most baby quilts, too."

"I hope to spoil him with quilts as he gets older so I hope you don't mind," she said with excitement. "There are so many fun designs. I'm glad you like it."

"Thank you so much, Pat," I said giving her a hug. "Sam and Sammy thank you too!"

# CHAPTER 60

I showered and dressed early for the big Christmas dinner so I could be helpful to Ella with the last minute preparations. Kip was in charge of setting up a bar on the sun porch so that was taken care of. I tried to work up my excitement once again, but hearing that Helen was so ill took the wind out of my sails. I would do everything I could to make her visit extra special!

I chose a long, red, wool skirt to go with the new black cashmere sweater Mother had given me last night. The sweater fit perfectly, and I knew I would enjoy wearing it often. The skirt I purchased when Sam and I were in Door County. I thought at the time it would be perfect for one of the many cocktail parties Sam and I would be attending in the future. The Hyde Side Boutique was my favorite shop in Fish Creek. I wondered if I would ever get back to that wonderful part of the country! I found the perfect pearl earrings and bracelet that complimented the pearl buttons on the sweater.

I looked in the mirror and asked out loud, "How do I look, Sam?" I knew he would approve. I touched up my make-up and off I went to pull off a spectacular Christmas dinner.

Sammy was still sleeping from his nap when I went downstairs. I went straight to the dining room to re-arrange the place cards at the dining room table since Helen and Pat had joined us. I wanted Jack to sit on the other side of the table next to my Mother. I didn't want anyone having any silly thoughts if I engaged him in conversation, so it was best if he sat far away.

The table looked simply gorgeous. My arrangement of a long, narrow centerpiece of holly intermingled with individual candle sticks was not only attractive, but would give everyone at the table a clear view of folks. I reminded myself to adjust the lighting with all the additional candles at dinner. There were going to be two different wines served with the courses, and each guest was going home with a bottle of the Red Devil wine that I had purchased at the auction. Would the reverend take one since he had a case of it at home as well?

I checked out the living room which was already overflowing with gifts under the tree, and the guests hadn't even started arriving. The tree looked almost miniature compared to the big one in the hall. I loved this tree, however, because I had it decorated in all gold and white ornaments. It glistened on its own even without the white lights. It, too, benefited from many ornaments I had picked out at the Christmas shop in Door County.

The house seemed extra quiet and smelled divine. I went to the den to put on Christmas music to play in the background. Last year at this time, I wouldn't have any of it played in the house with Sam gone for Christmas. I didn't see Ella buzzing around anywhere so I assumed she was upstairs dressing before Sammy woke up.

My cell phone went off, and I saw that it was Kevin.

"Anne, I don't know if you noticed, but the power went out in the greenhouses," Kevin announced somewhat stressed. I gasped. "I happen to come back to the office to pick up a hidden Christmas gift I had for Abbey and noticed it. I got the generators going right away, but we need to keep an eye out with these temperatures. Thankfully it wasn't out long when I discovered it, or we'd have some houses full of frozen vegetation. I haven't been able to reach Kip, but I figured you had him pretty busy today with going to the airport and all."

"Oh, thank goodness, Kevin," I praised him.

"We may be a little late this evening," he added. "Oh, before I forget, Abbey said that they found the guilty wire cutter!"

"They did. Who?" I asked in shock.

"It was the sweet little Sugar Plum Fairy!" he revealed. "Abbey can tell you more about it later."

"Thanks, Kevin," I said before hanging up.

I had thought the person I saw that night was a woman. Her motive had to be severe, to be so mean. Jack Frost and others said she was a strange one. I wondered how they discovered it was her.

Ella came looking for me. She looked absolutely magnificent. It was times like this, when she got dressed up, that I was absolutely confused about her age.

"Wow, you look so beautiful, Ella!" I complimented her. "I don't think I have ever seen that dress before."

"That's because it's new," she bragged as she turned around. "It's amazing what a new outfit can do for one's ego!" We both nodded and laughed.

"What needs to be done, or what can I do?" I innocently asked. She thought for a moment.

"I don't think it's too soon to fill the water glasses," she suggested. "I'm going to put on my apron and do the garlic mashed potatoes."

"Would you rather I do that, Ella?" I quickly offered. She looked at me with a shameful look.

"No offense Miss Anne, but you pour water much better than you make mashed potatoes," she teased. "It's going to be a challenge with such a large batch so this is where one's experience helps."

"Good thinking," I said as she went to the kitchen.

I was never going to be a great cook because being a great cook was never something I wanted to aspire to, nor was it on my bucket list. I was spoiled with good cooks in my life. I had a green thumb, and that was good enough for me!

# CHAPTER 61

I went to check on Sammy and found him amusing himself with his own chatter. This was a sign of him being in a good mood, for sure. I got his velvet outfit ready to dress him for the big occasion. Helen peeked in the room with interest as I proceeded with getting him changed.

"Say hello to Grandma," I said to Sammy as he smiled and clapped. "He's in a good mood so I hope he likes this outfit!"

"He is such a handsome fella," Helen admired.

"So what do you think?" I asked Helen as I stood Sammy on his feet to be admired.

"It's quite festive!" she said without hesitation. "You're having fun with him, Anne, aren't you?" I nodded with a smile.

"He makes me very happy," I glowed. "It makes life without Sam worth living." She looked at me with sadness as she paused.

"I'd like to go visit the grave site while I'm here, Anne," she announced with Sam on her mind. "I shouldn't be out in this blessed cold weather, but I must go."

"I totally understand, Helen," I consoled her. "If you want me to go, I will. I frequently take Sammy there in his stroller and we talk away to him. It's silly, I know, but it makes me feel better. One day I tried to get Sammy to say Da Da, and he refused to until we got to the car to leave. All of a sudden, he said quite loudly Da Da! Can you believe it?" She smiled with approval. "I show him his picture all the time, and I assure you, Helen, Sammy will grow up to know his father."

"Please do so, Anne, no matter where your life may take you," she advised with worry. "I wish I were here to tell him stories about his father. When I look at Sammy now, he is the spitting image of his father."

"What are you ladies up to?" asked Pat now joining us. "My, what a good looking chap we have here!"

"See if he will go with you, Pat," I encouraged. "I need to get downstairs to help Ella."

Pat reached to take him out of the crib. He looked at me as if to get permission.

"Perhaps he'd like to hear the story of The Three Bears," I suggested.

They both thought the idea was a splendid one, and I left the three of them in the room keeping my fingers crossed.

I went down to the sun porch to make sure there were still lights in the greenhouses. It appeared everything was fine, but I worried that with no break in the weather, we may have trouble ahead.

As I looked out the frigid patio room, nearly covered in ice, I couldn't help but wish my life away with hopes of spring to come. It was my favorite season. As much as I adored Christmas, spring was my thing. I supposed it was why I chose to have a flower shop. I could be among the living even in the dead of winter.

Ignoring Ella's business, I went about filling the water glasses in the dining room. The guests would all be arriving in an hour or so.

When Pat and Helen brought Sammy down to join us, Ella told them to put him in his chair for his Christmas dinner of smashed ham, peas, and peaches. Ella put a sizable bib on him, and Sammy started kicking in the excitement of food about to arrive. Pat and Helen were amused as they watched the whole process. The poem I wrote about Sammy's first Christmas came back to me as I watched. Little did he know what all this attention was about and how important these people would be in his life! Yes, little did he know!

# CHAPTER 62

When Kip arrived, I helped him set up the bar. I didn't realize we had so many spirits and beverages to offer. We carried the bottles of Red Devil to a small table by the front door so the guests could take one as they left. I took time to place red bows and holly on each bottle instead of using gift bags.

The first guest to arrive was Beverly looking so pretty in winter white. She was a bit stressed over the generators taking over in the greenhouses. She said she would make it her priority to check on them throughout the evening.

"That's not necessary!" I responded. "You are here as my guest tonight, and I want you to enjoy this amazing meal. There are enough of us here to deal with an emergency quickly if we have to."

"I know, but I feel like I really have no business sharing your family's Christmas dinner," she shyly noted. "My grandmother would be rolling over in her grave if she knew I was sharing your table!"

"Nonsense," I replied. "You are part of the Dickson family who does a great job for me. I want you all with me, especially this year. Everyone loves you, so just relax and enjoy! I wish I could have known your grandmother. She could tell me so much about this house as a cook here. Ella could write a book when she leaves here too I think!" We laughed.

"You are so generous, Anne," Beverly said as she went to get a cocktail.

I could see how she felt out of place as the only African-American here. It appeared that everyone loved her, or at least accepted her because she was so kind and a very hard worker. I wanted to keep her employed, but the greenhouses were so seasonal. I went to find her to explore an unexpected idea.

"Beverly, there are times at the flower shop that we could really use another person. This Forester funeral was a good example." She gave me a curious look. "Would you be interested in working in the shop? I will keep you on the payroll with a salary if you don't mind working where you are needed."

"But, Anne, I am not a floral designer," she cautioned.

"I know, but you are so good with flowers and so much of what we do at the shop can be learned," I explained. "There are a lot of grunt jobs, as we call them, where talent is certainly not required! Would you think about it?" She grinned from ear-to-ear.

"There's nothing to think about, Anne. I would be thrilled to help at the shop," she quickly responded.

"Good. We'll work out the details later," I added. "Consider it part of my Christmas present."

"Thank you so much, Anne, I can't believe this!" she said with a glow.

I had to answer the door, and Beverly floated away in disbelief.

"Why if it isn't my favorite niece and nephew!" I said greeting Sue accompanied with Devin. "Please come on in!"

"We're kind of early, I hope that's okay!" Sue apologized. "I think everyone was anxious and ready to go." Devin was holding Eli and it looked pretty natural for him.

"You are not the first to arrive. Let me have your coats," I said helping Mia take off her tiny coat.

"Devin, I'm so glad you could join us," I said making him feel welcome. "The bar is in the sun porch or Kip may come by and take your order. Mia and Eli look adorable in their Christmas outfits!"

"Well, for the first time in my life, I helped this little fella in his suit. He does look pretty dapper, don't you think?" Devin bragged. Sue laughed.

"I think Devin was overwhelmed by what one goes through to get everyone dressed and then coats put on," Sue described. "We have this bag of gifts. Should we put them under this big tree or the one in the living room?"

"The living room is already overflowing, but that's where we'll end up," I informed her. "My goodness, you did overdo, Sue!"

"Well, it's Christmas and I never get tired of shopping when I get the time." Sue took Mia's hand and Devin and Eli followed to join the others. They sure did look like a readymade family.

Mother and Harry were the next to arrive, and Mother's coat was hardly off before she made a quick dash to the kitchen to make the gravy. Harry shook his head and commented how it was impossible to keep up with her. Ella came by and offered him one of the appetizers she was about to take to the study. Ella and Harry had that special history so it was pretty nice that they could share the holiday together at my house. Harry assisted her with a platter full of shrimp and cocktail sauce that looked heavenly. I was getting hungry. I managed to take one before I once again had to answer the door.

# CHAPTER 63

Look who's here! Merry Christmas!" I said to Aunt Julia, Uncle Jim, Sarah, and the mysterious Jack Fletcher, who was right behind them. Jack was grinning ear to ear knowing he felt lucky to be included in this clan. I made some small talk about the relentless snow flurries that so far hadn't amounted to anything.

"Where's Sammy?" Sarah immediately asked. "I got a new camera for Christmas, and want to take some pictures of him!"

"Oh, that would be wonderful!" I responded. "He's in a pretty good mood. Be sure to get some of him with his Grandmother Dickson." She nodded running off.

"I think Sally drove up right behind us," Aunt Julia said taking off her coat. "Are you sure we won't slide down that hill when we get ready to leave?"

"No, I can't guarantee that!" I teased. "What I can guarantee you is that you will have a marvelous Christmas dinner!" She cheered.

"You look beautiful," Jack finally said. "Red seems to follow you. I can see why you have a fondness for the color." I blushed as I took his coat.

"I couldn't agree more!" chimed in Uncle Jim.

"I'm beginning to feel a little guilty being at this lovely place again this holiday," Jack said softly.

I didn't get a chance to answer when Pat came up to me with Sammy in her arms.

"He was getting pretty persistent in asking for his mama, so I thought I'd better help him out," Pat revealed.

"Pat, this is Reverend Jack Fletcher from our church." I introduced Jack as I took Sammy from her arms. "This is Pat, Sam's sister and Auntie of this little guy."

"Please call me Jack," he insisted. She looked amused and nodded her head as she shook his hand.

"Well, I have looked forward to meeting this little man," Jack said smiling. "You are as handsome as your mother described."

To my surprise Sammy smiled and reached out his arms as if he wanted Jack to take him. Jack immediately responded which shocked me.

"Oh Jack, he is teething and can be a handful, I'm afraid," I said in my awkwardness. "I can't believe he responded this way. He has been such a mama's boy lately."

"I can't believe it either," Pat contributed. "I think he's probably glad to be with another man for a change. There are a lot of females in this house!" We laughed and then Helen came up to join us.

"He seems to know you, Reverend," Helen boldly stated.

"This is Sammy's Grandmother Dickson, and this is Reverend Fletcher," I quickly introduced them. They acknowledged each other very quickly.

"Actually, I happen to be pretty good with children," Jack added. "I have some nieces and nephews, and I occasionally teach a Sunday school class of little ones." Helen seemed impressed, but little did she know the reverend's intentions.

Jack then turned and carried Sammy into the study where Uncle Jim had joined the others. I let in Sally who had white flakes on the top of her hair.

"What's this, a new hair cut?" I claimed. "I especially like the touches of white!"

"Yup, just got it," she said taking off her coat. "Do you like it? I'll have you know that I am wearing a skirt for the first time in ages, and I'm freezing cold! What I wouldn't do for you!" We both chuckled as I watched her shiver from the cold.

"Why the changes and why wear less clothes on a day like today?" I jested.

"It's Christmas, that's why!" she cheerfully said. Well, I thought it quite interesting but dropped the subject.

I told her where the bar was and tried to determine who was still missing. It was Abbey and Kevin. It made me think to see if the lights were still on in the greenhouses. I walked out to the sun porch where Kip handed me a glass of Red Devil and told me to relax and not worry.

"Things are going very well," Ella reassured me. "I see Sammy found a new friend!" She winked.

"Stop it, Ella," I quickly responded. "Sammy actually reached for him, so what was I supposed to do?"

"You mean like when the reverend cut in on you to dance at the party?" she teased back.

"Ella, you are going to be in a lot of trouble, if you don't stop," I warned. She shrugged her shoulders and smiled.

Beverly came to the porch to tell me that Kevin and Abbey had just arrived. I gladly left with my glass of wine to greet them.

# CHAPTER 64

I t's about time you guys got here!" I scolded.

"Sorry, we had a lot going on today," Kevin said with frustration. "It looks like the greenhouses are holding, so that's good news."

"I sure hope so," I added with concern. "Beverly and I keep checking."

"Not to worry, Anne," Kevin said as he took their coats and walked away. I supposed he knew Abbey and I would want to talk.

"I heard we caught the bad guy," I said as a lead in to her news. She took a deep breath.

"Yes, can we talk now?" Abbey asked. I nodded.

"Well, the characters were asked to gather after the shops closed to turn in their costumes and accessories," she began. "I was warned by the guy who plays Scrooge that our little fairy was quite upset with us for refusing to hire her for next year.

She told him she was not going to turn in her costume, and he thought I ought to know." I shuddered what that might mean.

"It made me furious, and I decided I would confront her head on at her locker. She was there pretending to take care of business with her locker open, but she wasn't removing any parts of her costume. I went up to her and told her that everything had to remain including her wand. She looked at me with those hateful dark eyes of hers and threw the wand in the locker. I reached in to take hold of the wand, and I saw some merchandise that we had for sale in the tent. I was about to question whether those items were paid for when I also saw a pair of wire clippers. What would she be doing with those? When she saw me pick them up, she knew she blew it."

"Oh no, then what?" I asked with fright.

"She saw the look on my face, and I said, 'Wire cutters? How handy!' I didn't accuse her of anything. I didn't have to. She totally lost it and started cursing. Then with a nasty laugh, she bragged about costing the poor committee all the money for lights. She called us a bunch of hypocrites and said we would pay a price for her losing her job."

"We'll pay the price?" I asked with anger. "What price will she have to pay?"

"By this time George and others came around when they heard her yelling," she went on. "George got in her face, pushing me back, thinking I may get hit I guess. He warned her again to leave all of her costume and that he was reporting her to the police. When she tried to make light of his threat, he said she'd better not leave town because we would be pressing charges."

"So then what?" I said in disbelief.

"She threw her wings on the ground and turned around and left wearing most of her costume," Abbey reported. "It was a scene to behold. Everyone was stunned, and there wasn't much we could do about keeping her there."

"So you think she stole the merchandise as well?" I wondered.

"Oh, yes, I would bet on it," Abbey nodded. "You know, the more I think about all this, I think she rather enjoyed getting caught. Anyway, it's out of my hands. George said he'd handle it. I'm sure I'll have to testify down the line."

"Well, the good news is that more time didn't go on, and we found out who it is!" I consoled her. "I bet you are ready for a drink, aren't you? I give you a lot of credit for confronting her, Abbey."

"I am ready to relax and enjoy a part of Christmas where I don't have to work." Abbey added in relief.

"Now, you know, if you own a Christmas shop, I don't think you could escape working," I teased.

"I hear you, but I don't think I have to worry about that happening!"

As we joined the others in the study enjoying themselves, I was surprised to see that Jack was still entertaining Sammy. It was an unsettling picture for me as I watched my son with a man that was not his father. I went over to rescue him.

"I think we'll let Sammy have a little cookie to eat before he goes up to bed," I said relieving Jack. "Thanks for your help."

"My pleasure," he said with a big grin. "He is really a bright little guy, and he wants to keep moving, doesn't he? Does he get that from you?" I had to snicker.

"He probably does," I said curtly as I walked away into the kitchen.

I put Sammy into his chair, and Kip teased me about putting a cocktail into Sammy's sippy cup.

"There are times I have thought of that myself, Kip," I teased. "I think he's had enough attention here in the last hours to wear him out so hopefully he'll be ready to crash soon!"

# CHAPTER 65

Mother and Ella were showing signs of nervousness as we got closer to meal time. Timing was always everything in a big meal like this. I hurried Sammy's treat before taking him upstairs. As I was getting on his flannel pajamas, Pat joined me in the nursery. She happily stared at the challenge of getting Sammy to cooperate.

"He and the reverend seemed to get along quite nicely, Anne," Pat remarked. "Does he come visit here often? Sammy acts like he's familiar with him." I wasn't sure how to answer.

"It's only the second time he's been here and the first time he's met Sammy," I shared. "We have some mutual friends, and when I ran into him at the Barrister Funeral home, he mentioned he didn't have plans for Christmas. He just arrived here so going home to his family wasn't feasible. So, I invited him which I knew would delight my Mother. She thinks he's pretty great."

"And you?" she asked suggestively.

"He's very nice," I calmly answered.

"He's single, right?" she continued to pry.

"So I hear," I answered. Where was she going with this?

"I know it's none of my business, Anne, but have you thought about seeing someone?" She asked. "Perhaps you already have and it's none of my business. Knowing my brother, he would not want you to go through life alone." I was shocked by her openness.

"I'm not alone," I sharply responded. "I have Sammy and a whole shop family that I can hardly keep up with."

"Okay, I get it, but when I saw how happy Sammy was with a man's attention, I couldn't help but think you might, and should, have that for both of you!" I remained speechless. We quietly walked out of the room into the hallway as I looked for the right words to say.

"I know you mean well, Pat, and your opinion means a lot, but I still miss Sam terribly each and every day. I cannot look at Sammy without thinking of him. I had a perfect marriage!" Pat shook her head.

"Now don't put that workaholic brother of mine on a pedestal, Anne," she declared. "You both were lucky for sure, and no one will ever replace him, but perfect he was not! You should never close the door on what else life has to offer you." *Interesting response in many ways*, I thought.

"Thanks, Pat," I said with affection. "I appreciate the advice. We started going down the stairs. "I can't believe Sammy hasn't made a peep knowing what's going on down here."

We went to the kitchen and received the word from Ella that dinner was about to be served and we should tell everyone to come into the dining room and look for their place card.

"I'm starved!" yelled Sarah rushing past me.

"We have a youth chair for Mia," I told Sue.

"Eli's half asleep. After we pray I'll put him down in the maid's room as you suggested," added Sue.

"Whatever you like," I said to appease her. "So far, so good with Sammy."

Mother and Ella remained unseated as everyone else took their chairs. The side board had been filled with extra side dishes as the main meal courses would be placed on the table for a family style meal.

As I took my chair, I glanced next to me and saw Jack sitting there. I specifically placed him on the other side of the table. I looked at his place card to confirm his spot, and it said his name. How could this happen? I took a deep breath deciding whether to say anything or not. They were all looking at me to start the prayer, so I began.

"For the first time, our family has in attendance a man of the cloth, so I would like to ask the Reverend Jack Fletcher if he would lead us in prayer today." He looked surprised.

"Very well, Anne, I'd be honored," Jack responded.

"Let us join hands," he began. Oh, no, he took my hand.

"Dear Father, we'd like to honor you on this day of your birth with love and thanks. We are blessed to share this day with friends and family, as we wish everyone a Merry Christmas and a blessed New Year! In your name we share this joy! Amen." I quickly took my hand away.

"Thank you," I said politely. "Before we begin this delightful meal, I'd like to make a toast to Ella who can make great meals and events just happen! Cheers to Ella!" Everyone raised their glasses and cheered.

"Well, then I'd like to have the floor to toast our lovely hostess, Anne Dickson!" Uncle Jim announced. "She has a way of bringing joy and love to this extended family, and we all appreciate sharing this Christmas dinner with you!" Everyone cheered even louder!

"Sorry to delay this meal even longer," Pat chimed in. "From the Dickson side of the family, I'd like to make a toast to the new heir that will hopefully carry on this family tradition. Here's to my nephew and Anne's son, Sammy!" I had to laugh and hoped that somehow Sammy would hear us as we cheered. I grinned from ear to ear at their generous toasts.

"Ella, you better interrupt this madness, or the food will be cold and we'll all be full of wine!" I teased. Everyone agreed.

I joined Ella and Mother in the kitchen as we brought food to the table. Raves and sighs of hunger came from every direction.

"I was surprised to see you sitting next to me," I said as I sat down. He looked at me oddly.

"I just sat where my place card was," he defended. "Should I have been seated elsewhere?"

"I placed you at the end of the table," I honestly reported. "Someone rearranged the place cards."

"Are you offended?" he asked with some flirtation. "Does it make you uncomfortable?" I was now blushing, and he saw it.

"I just don't want anyone getting any ideas, that's all," I said softly so no one else could hear. Sue was sitting on the other side of me, and I couldn't tell if she was listening.

"What idea would that be?" he coyly asked. I coughed not knowing how to respond. I decided to ignore his question and make an announcement.

"While you are enjoying your meal, I'd like you to be thinking about a favorite Christmas memory you'd like to share with us when we start dessert," I requested. "Mother and Ella, the meal is simply delicious." The chatter became louder. It was certainly of mix of happy noise!

# CHAPTER 66

Thankfully, Jack did a fair amount of conversing with Uncle Jim who was sitting on the other side of him. The more wine Uncle Jim had, the more talkative he became. Perhaps he was in some way confessing things to Jack that he wouldn't be telling anyone else. Occasionally, Jack would make comments about the food, and said that he wished he had learned to cook more from his mother who must have been a great cook.

"So what occupation did your parents have?" I asked in general, when there was a lull in the conversation.

"Both of them were school teachers," he proudly stated. "That's how they met. They were both teaching at a grade school when they were young. My father went on to teach history in middle school, but my mother continued to stay with her first graders, whom she loved."

"How nice," I said picturing them in my mind.

"I'm really stuffed," Jack said with a stretch. "I will do my best, however, to enjoy some of that amazing cake I see over there."

"I hope so! The cake has quite a reputation! Nick's Bakery makes this cake for most special occasions in Colebridge! When we get things cleared away, we'll start our stories while we eat our cake." I saw most everyone had finished so I explained the next part of the plan.

"Please continue eating if you are not finished," I began to explain. "While Ella is removing some of the plates, I'm going to cut the cake so we can get started telling our stories."

"I'm finished, Anne. I can help Ella," offered Sarah.

"That would be great, Sarah," I said, nodding with approval.

"I'll be happy to refill the coffee and tea, Anne," offered Beverly. With that everyone thought they were ready for dessert.

"Great, but everyone else please stay put," I instructed. "This will just take a minute."

"I'll have more of the red wine," Uncle Jim said to Beverly. She quickly obliged.

I started slicing the heavenly chocolate cake and putting it on plates. Kip leaned back from his chair and whispered, "Just so you know, it's really starting to snow now. It's covering the ground."

"I didn't hear that," I whispered back. He knew there wasn't anything that was going to spoil this Christmas dinner, and I didn't want everyone to look out the window to spoil the evening.

When Ella, Sarah, and I started passing out the plates of cake, we heard sounds of delight and groans of discomfort as some wondered how to make room for the cake.

"Okay, I will begin the story telling by telling you one of my simple but delightful Christmas memories," I said, taking a deep breath. Mother was listening intently. "You see, I loved pretty clothes—still do. Since our Sunday school classes always had a Christmas program every year, I got to pick out my new dress for the occasion. One year I chose a red and black sparkling taffeta dress that made noise when I moved. When I recited my verse, it made a swishing noise which made the other kids giggle. That's my memory. See it doesn't have to be anything monumental!" Everyone politely snickered at the cute report and applauded. Mother shook her head like she had heard this for the first time.

"Well heck, I can do this," announced Kip. "I'll go next. My memory is honestly a bummer, but there isn't a Christmas that doesn't go by that I don't think of this little trick that I played on my brother. My younger brother Ken and I were close in age so we got a lot of the same presents. One year I hid a couple of his gifts and told him that Santa only brought him a shirt. He saw all my presents and started crying. It took a while for my parents to catch on to what had happened. Needless to say, I was grounded that Christmas." We all shook our heads.

"I will remember that, Kip," I joked. "I didn't know you were a bad boy!" Everyone chuckled.

"Hey, mine is quick and simple," jumped in Kevin. "I remember this like it was yesterday. I had a paper route most of my boyhood, and I had the worst excuse for a bike you ever saw. Well, finally one year, there was a brand-new red bike by the Christmas tree. I thought I was in heaven! After that, I had more friends, and I thought I was king of the road." We all laughed and applauded. "Now, I have some really good Christmas news I'd like to tell you all while I have your attention." We all stopped talking and gave him our full attention. "I've asked Abbey to marry me in the coming year! She said yes, and we are now officially engaged!" I got chills all over. The sighs and applause were overwhelming. I wanted to cry for both of them. Abbey was blushing.

"Abbey, what do you have to say about that?" I jokingly asked.

"I found my soulmate and I'm not letting him get away!" Abbey bragged. "My ring is amazing and I am totally happy!" Everyone applauded, and Kevin and Abbey gave each other a kiss. All eyes were on the ring she proudly showed on her left hand.

"Congratulations to you both," I acknowledged. "Abbey, would you like to share your best memory—besides this one of course?"

"Sure," she said trying to gather her thoughts. "I grew up in New York in an apartment building over a meat and sausage shop. The owner became like family to us over the years, and on Christmas Eve, he would give everyone in the building some of his homemade sausage. You could smell it everywhere in the neighborhood on Christmas morning. I've never forgotten that pleasant smell—even today!" We all smiled and applauded. *Sometimes the little things are the best memories*, I told myself.

"Abbey, I want to give you another Christmas memory, if you can possible handle another one besides that darling engagement ring," I announced. The room once again went perfectly still. "I debated on whether to do this at the dinner table, but we'll discuss that later," I said. "Abbey did a magnificent job managing the Mistletoe Market this year for our Merchants Group. Ever since I met Abbey, I saw her creative talents and love for Christmas. Her dream is to have her own Christmas Shop on Main Street which we really need. I am going to financially make that happen for her when she is ready to make the plunge." All were silent as they looked at Abbey.

"What?" Abbey said in shock. "I don't know what that exactly means, but, Anne, that is way too generous to imagine!" Kevin gave her a kiss and shook his head in disbelief!

"I think it would be a wise investment for you and Main Street," I bragged. "I don't want dollars to keep you from moving forward on your dream!" Abbey got up out of her chair and came over to hug me. Tears were flowing down her face.

# CHAPTER 67

"Wow, this is a lot of good stuff," I admitted as I tried to gather my emotions. "Mother, why don't you go next?" She shrugged her shoulders like she was embarrassed.

"When Anne told me to think about a Christmas memory, it was hard to think of just one," she admitted. "I did get engaged to her father on Christmas Eve, and I don't know if Anne even knew that!" She paused and continued. "As a child however, Christmas Eve was the highlight of the year without question. Back then, we didn't have our presents wrapped. We would come into the parlor after the Christmas Eve church service, and there were our gifts. I got a new doll every year, so that was what I would be looking for. Now, I'm not sure girls play with dolls anymore, but those moments were priceless." Everyone smiled and applauded once more. No, I didn't know the engagement story.

"Anne, this is such a nice touch to the evening," Jack whispered as he leaned over to me. He had a nice soapy smell that I remembered from before.

"I want it to be special for each one of you," I said before continuing. "Ella, I know yours is a bit sad, but memories are about life and death at Christmas." She took a deep breath.

"It sure is," Ella began. "I lost my father on Christmas Day. He wasn't feeling well that morning so we all went to church without him. When we got home, we found him in his favorite chair. Mother checked on him and when she came to tell us, she said, 'Your Daddy was a gift to Jesus this Christmas.'" Ella tried to keep from crying as everyone near her at the table expressed their sorrow.

"Your mother sounds like she was a good, Christian woman," added Jack. Ella nodded.

Just then, the lights began to flicker on and off. We all looked at each other in alarm. Kevin and Kip jumped out of their seats immediately to go to the sun porch where they could check the condition of the greenhouses. They came back to the table to announce that the lights were still on, but that they better get over to the office to check on the generator in case we would lose power. Abbey insisted on going with them.

"Do you need my help?" asked Beverly. They told her to stay put, but there was no way Kevin was going without Abbey.

"Please everyone, those guys have everything under control, trust me," I said assuring everyone to stay seated. "I have a whole cabinet full of candles in case the lights go out again. We have twelve candles right here on this table, so we'll be fine. Let's see, who would like to go next?"

"I'll go," Sarah said with enthusiasm. "I mostly think about Grandmother Davis at Christmas because she always gave me money. No one else ever did, and I loved it. After she died, I still continued to get a card with twenty dollars in it. So this morning, under the tree as always, was my twenty dollars. Mom's not talking so I guess Grandmother is still taking care of me." Everyone laughed and then clapped. She's right, Grandmother Davis is still around doing her business. I'd never heard this story before.

Just then, total darkness came over the room except for our lit candles. Everyone had sounds of concern.

"Stay put everyone," I shouted calmly. "Ella and I will get the candles. We'll continue with our stories. No one can go anywhere without light, so stay put and keep enjoying your dessert."

"I'll help," said Jack who was getting out of his chair.

"No, please stay and keep this crowd calm until we can light a few of the candles," I instructed him.

"Yes, ma'am," Jack said with obedience in his voice.

Small talk erupted at the table as Ella and I easily found our way to the kitchen cabinet. Sam had stocked enough candles to light the whole house. We put them on a tray which we carried to the dining room. Ella lit two larger candles on the buffet, and we went back to our seats.

"This is quite romantic and spiritual, don't you think?" I said in a teasing voice. Jack shook his head in disbelief of my ability to adjust to the situation.

"I would like the stories to continue," said Jack in a controlling manner. "This is very special, Anne."

"So we will," I responded. "Uncle Jim, while you are still sober, I think you should tell us something." Everyone laughed, but Aunt Julia gave him a dirty look.

"Hey, it's Christmas," he said. "Okay, Julia will remember this one. Remember the year we told my mom and dad we were expecting a baby? We didn't have a dime to our name so to speak. I wasn't sure if we destroyed their Christmas or made their Christmas, but my father groaned when we told him and my mother cried. Julia and I just looked at each other. Of course, Sarah turned out to be the apple of their eye, but that Christmas day was not the best memory for me." Sarah laughed looking now at her mom.

"Okay, Jim," Aunt Julia followed. "We did get a better reaction from MY folks when we told them at the Christmas dinner table. Mother grinned and dad asked me to pass the mashed potatoes." Everyone chuckled. "When I was helping my mother with the dishes later, she said. 'Well, you may as well start a family because I don't think you can hold a real job.'" Everyone now softly groaned in sympathy. "That's okay. That was my mom. The glass was always half empty, never half full." I looked at Mother who was not pleased with Aunt Julia for sharing the story. They all gave her a small round of applause.

The lights were still out, but everyone seemed to be getting comfortable with the atmosphere. Ella was refilling their cups and pouring more wine.

"I can identify with Kip's story," began Beverly. "I had a lot of siblings growing up with no money to speak of, but on Christmas my folks would make sure we had something to open on Christmas morning. It took us a while to realize it, but mom and dad never had a gift to open. So one year, we pooled our pennies, nickels, and dimes to get our parents each a gift. We got my mom a cheap bottle of perfume at the dime store and my dad a pair of socks. I can still see the look on their faces when we gave them their gifts. Let me tell you all, it was the year I realized it was a better gift to give than to receive. We all cherish that moment to this day!" There were some watery eyes around the room, and Beverly's story drew the biggest applause.

"That is a precious Christmas story, Beverly," I said with tears.

"I loved it as well, Beverly," chimed in Jack. "I may ask to borrow that for a sermon."

Just then, Kip came in from the outside and entered the dining room. We waited to hear what he had to say.

"Just so you know, you better all get comfortable here because no one will be able to go anywhere tonight." he announced shivering. "The snow is getting deeper, and there's no way anyone will get down that hill with the ice underneath the snow. The generator is still going strong for the greenhouses, but we have nothing to keep this house warm, other than the gas stoves, so you can all be thinking of spending the night!" Gasps of surprise were voiced around the table. I had to think quickly.

"Cool," said Sarah.

"Don't worry everyone," I said to calm the concerns. "We have lots of beds and quilts in this big house, and we will stay warm. Let's stay together at this table while we decide future arrangements." The looks of concern were not going to rattle me. Besides, the lights could come back on anytime.

"Thanks, Kip," I said. "Merry Christmas to you, too!" Everyone laughed.

"Hey we have a very hot pot-belly stove over at the office that puts out tons of heat, so if anyone gets too cold here, you're welcome to join us over there!"

# CHAPTER 68

I think we are lucky to be right where we are," Harry announced to calm everyone. "I will continue the stories, Anne." Everyone applauded for him to continue. "I am probably the oldest one here, so a lot of my memories were when I was in the service. I found out pretty quickly that girls love a man in a uniform!" Harry winked at Mother as she blushed. "I got a lot of mail from girlfriends here and there around Christmas, but I knew no one would really be there when I'd return home. I'll never forget the lonely gut feeling I had being away from home at Christmas. I have never taken that holiday for granted again as I celebrate each Christmas. This one couldn't be happier here with my dear Sylvia." Now everyone looked adoringly at Mother and Harry as he kissed her on the cheek.

"Why thank you, sweetie," Mother said with affection in her eyes.

"That was wonderful, Harry," I responded sniffling. "These memories are too much!" I glanced over at Jack who was taking it all in.

"Well, my Christmases were not all so great either, Harry," added Sally who finally came forward to speak. "I was sometimes away at boarding school for Christmas. Before I knew it, I came home to parents who announced they were getting divorced." Everyone got terribly quiet again. "When I did come home after that, the stress was awful trying to appease each one of them with where I was going to spend Christmas. One year, I said to myself, 'I've had enough.' I stayed home and invited all my lonely friends over and cooked a big meal. The food was pretty horrible when I think back on it." She had to stop and laugh. "We gave each other silly gifts, and it was one of the best Christmases I can remember!" Everyone applauded and commented about similar experiences.

"I know Mother is still thinking so I'll go ahead," said Pat taking a deep breath. "I know she's wondering if I will tell this story, but it is hard not to recall it immediately. I was only 6 years old when our Christmas tree caught on fire from those old fashion bubbly candles." Everyone gasped as they looked toward Helen. "My brave father picked up the tree and took it outdoors so it wouldn't burn the house down. I just remember running around screaming. My dad had burns all along his arm." Silence at the horror of it all did not get applause.

"Oh, Pat, I remember Sam telling me a bit about that," I said sadly. "How awful!"

"It could have been much worse," chimed in Helen. "It was a night to remember, for sure! On a happier note, I must tell a little story about my little Sam. Christmas morning couldn't come early enough for him every year. One year, he got up very early and decided to open all the presents. I think he was around 7 years old. Some of the presents were his and some were not of course. He knew he was going to get in trouble, so he tried to put all the presents back like they were before he opened them. The only person more upset than his sisters was his father. I'm sure if Sam were here, he would tell you that he would never forget that Christmas." I had to laugh as did Pat.

"Sue, I don't think we've heard from you!" I said looking at her with a smile.

"Well, as I grew up, we usually went to Grandmother Elliot's house for Christmas. She lived a couple hours away until she went into a nursing home. We would pack up gifts, Christmas cookies, and always mother's cranberry relish that she made each year. One year, my dad was carrying the relish, and as he started to put it in the car, he accidently spilled it all over the back seat. My mom really let him have it, and they didn't speak for days. It was not a fun Christmas, and to this day, my mom won't talk about it. I can't eat cranberries on this day myself because I think of them all over me and the back seat!" Everyone laughed and gave her some applause.

"I didn't grow up around here, either," Devin finally said. "I grew up in Minnesota where we had snow most of the year it seemed. We lived out in the country to make it even worse. I had a sled that I used and used until it fell apart beyond repair. The thought of going through another long winter without my sled was depressing. When Christmas came, we were used to getting only one gift which was usually handmade by mom or dad. Well, I must have been a really good boy that year because there was a brand new red sled under the Christmas tree. It was not homemade but store bought. We all went crazy. I have it to this day!"

"I can imagine the wear and tear you put on that thing!" joked Uncle Jim. "I loved my sled." Everyone clapped and told him how much they enjoyed his story.

"I think we've heard from almost everyone except the reverend," I revealed. He looked at me and grinned.

"I've been thinking, and it's hard to top these wonderful moments you all have been sharing tonight," he said slowly giving it some thought. "This was a brilliant idea, Anne. Christmas is so many things to so many people under different circumstances. I grew up in the snow of Madison, Wisconsin, so I know what you mean about having something mundane to get you through the long winters. For me it was ice skates. One year I got new ones so that was pretty memorable. We belonged to a very small church. In fact, my mother still attends that very church. On Christmas Eve, they gave every child a large orange and a small box of chocolate candy. For some of the less fortunate; that was their total Christmas years ago. They still practice the same tradition today and, of course, the young folks joke about it, but they don't turn it down!" Everyone laughed. "Today, it's a different world with all the electronics and expensive gifts they receive at home. Frankly, I think if they stopped the tradition, many would be disappointed."

"That's pretty special and sends a nice message, doesn't it?" I asked as they all applauded.

"Thanks so much to all of you for sharing these memories with us," I acknowledged. "I know Santa was here earlier today and brought us some  presents to open. Sammy is sound asleep, and I happen to know most of the gifts are for him. We'll put those aside for him to open tomorrow. It may start getting chilly in here so let's all go to the fire in the study to open gifts. We have plenty of warm refreshments. We also have lots of quilts and blankets to get comfortable with. Ella and I will make sure you'll be comfortable for the rest of the evening. Some of you probably need to make some phone calls, but rest assured, I am not letting any of you leave this evening."

Everyone got up and started chatting about the arrangement. Many went to look out the windows. We were now having the white Christmas so many had been asking for.

# CHAPTER 69

Everyone rushed from the table to find a nice warm spot near the fire in the study. Jack and I took some of the candles Ella provided and started lighting and placing them throughout the house. I took one of the candles up the stairs to check on Sammy. He continued to sleep on without a care in the world. I took the lily quilt Grandmother Davis had provided for him and added it to his blanket to keep him warm. This room would be cooling off quickly.

I placed candles along the hallway so everyone could find their room, and made sure each bedroom had one or more candles. Ella then joined me as we pulled blankets, quilts, and pillows from the hall closets.

"I think we really do have enough beds and couches if everyone agrees to my suggestions," Ella noted. "All the beds are made up except for the yellow room. I've never made up those twin beds before but will take care of it right away.

Julia and Sarah can then have that room."

"Oh dear, where will we put Sue and Devin?" I asked concerned.

"Mia and Eli are already asleep in the maid's room downstairs, so I told Devin to unfold the roll away in there," Ella reported. "No sense in moving those kids, plus they will want to be with them."

"You're right," I responded. "I told Sally she could share my room. I have slept in Sam's recliner before and it's quite comfy! Oh dear! What are we going to do about the reverend?" Ella laughed. She must have thought of all this during dinner.

"He can take the couch in the study if he hasn't already," she joked. "I told Beverly to take the couch in my room."

"Ella, you're something!" I said in disbelief.

"Take this bedding down for the two couches while I make up those twin beds," Ella instructed.

"Yes, ma'am," I said taking the bedding from her. Mother was waiting for me at the bottom of the stairs.

"Honey, you need to join us in the study to open your gifts," she instructed me just like when I was little and strayed away.

"Okay," I said joining her. "We have the green room prepared when you and Harry are ready to go up." She nodded.

"Sit here," Sue called out as she pointed to the foot stool near her. She handed me a gift, which I received with a big smile. I almost forgot it was Christmas.

When I opened the lovely box, it was a pearl and shell necklace and earrings. I knew that she probably made it because I knew she was taking some classes. I held it up for everyone to admire.

"This is very clever, Sue. I bet you made this, right?" She proudly nodded.

"You must go in that shop near the Water Wheel called String and Stone," she told me. "They have so many unique things, and I really love their classes."

"I love it! Thank you for adding the pearls which I seem to gravitate to," I said in response. Everyone voiced their compliments, including Jack. He was sitting on the floor looking quite comfy by the fire. He certainly had no problem fitting in with this crowd.

"I brought you something, too," cried Sarah. She handed me a gift bag. "Promise not to laugh." I smiled.

I pulled out a red knitted scarf. I remembered Aunt Julia telling me that Sarah had picked up knitting with her friends.

"Oh, Sarah, its darling! You picked my favorite color!" I flattered her. She grinned. "I can't do this. Was it hard to do?"

"It's the basic," she responded with shyness. "It goes pretty fast. I made Mom one in green to match her coat."

"Thanks so much," I said giving her a hug. "I'm glad you like the button accessories I got for you!"

"Are you kidding?" Sarah came back in response. "I can't wait to wear the necklace. I love that button shop. I go in there quite a bit!"

# CHAPTER 70

O h, Anne, I brought you something to wear as well," called out Aunt Julia as she handed me a gift bag. "Jim has something for you, too, but I don't know where he is. Have you seen him?"

"Good question. No, I haven't," I responded.

"He's probably helping himself to some more wine in the kitchen," Aunt Julia said in disgust. "Sarah, see if you can find him."

"Well, the good news is he won't be driving home this evening," I said in jest.

I opened the gift bag, and it was a black and white scarf with writing on it. When I looked closer, I saw it was all Jane Austen quotes.

"Oh, Aunt Julia! How cool is this?" I exclaimed. "Are you selling these in the shop? I've never seen anything like this before."

"It is very lovely, Julia," admired Pat who was also a Jane Austen fan. Aunt Julia smiled knowing she did very well in choosing my gift.

"You should see her shop, Pat, it's very clever," I bragged. "She has a lot of Jane Austen items for sale."

"I owe a lot of it to you, Anne, but thanks," Aunt Julia blushed.

"Here's something from Mother and me," said Pat handing me a larger bag which I assumed was the baby quilt she mentioned.

"Isn't anyone else getting gifts?" I asked looking about the room because I felt guilty.

"Mother and I exchanged earlier," Pat confessed. "We both had to laugh because we made this solemn promise that the trip was going to be our only gift to each other."

"That's sweet of you two, but it is Christmas," I teased. "Surprises are likely to happen!"

I pulled the tissue away from the bag and pulled out a very special quilt for Sammy. It was indeed a pictorial of The Three Bears which was Sam's favorite book as a child. I could tell there was so much work put into each detail. I stood up to show everyone else in the room. Everyone applauded and offered compliments.

"Thank you both so much," I said tearing up, as I thought of Sam. "Sammy will love this! Between his train quilt and this, I'd say my Sammy was a pretty lucky grandson, wouldn't you?" Everyone agreed.

Out of the corner of my eye I watched how Jack was following all the activity. I didn't want to look directly at him. Devin seemed to be the only other man giving Jack some attention. Sue kept running back and forth to the bedroom where Eli and Mia were asleep.

I went across the room to sit by Helen for a while. She looked very tired. I asked her if she was okay and she commented that she would be going upstairs shortly.

"I really love the quilt, Helen, and I appreciate you making this trip," I said as I put my arm around her. "You are surprising me with how well you are doing."

"I must keep going you see," she said looking away as if she were going to cry. "I had to see this precious grandson of mine and know how you were doing. I see keeping busy seems to be your salvation."

"It is, Helen," I admitted. "I grab Sammy in my arms when I feel the least bit sorry for myself."

Pat now joined us and suggested they turn in for the evening.

"I have a little something else for the two of you," I said reaching for their gifts.

"Oh, Anne, you have done enough by paying for our flight," voiced Helen.

I handed them each the same gift, and Pat was the first to open hers. She pulled out a framed photo of Sammy. I told Helen she had one just like it. They appeared to be thrilled.

"Here are two more exactly alike I'm afraid," I said as I handed them a tiny gift-wrapped box.

Helen was the first to open the sterling silver charm bracelet that had a charm for Sammy's birthday and then one for Sam Dickson Properties. They both seemed totally surprised and pleased as we shared hugs and kisses.

"I'm glad you like them, but it doesn't compare to the time you put into Sammy's quilt," I noted.

"I'd like you and Sammy to visit us soon. Could you possibly do that?" Helen begged.

"I will certainly give that some thought," I said with confidence.

"With that note Mother, we need to get on up to bed and get warm," Pat encouraged. "The dinner was magnificent, Anne. Thank you so much. Little did we know we'd have the surprise of a white Christmas to boot!"

I followed them to the staircase giving them each a candle. After they got to the top, I walked down to the kitchen and then to the sun porch to make sure the lights were still on in the greenhouses. The rooms were quite chilly compared to the warmer rooms I'd just left. I couldn't help but put in a little prayer request that the lights and heat would be on soon. I didn't envision any sleep for me until all was well. What an unpredictable night this turned into!

# CHAPTER 71

I jumped when my cell went off. It was Kip reassuring me they were all warm and toasty around the fire in the Brody office and they had no intention of leaving there. He noted that they had found extra blankets in their cars to help keep them warm. He thought the snow would let up soon according to the forecast, which meant that the snow plow would show up in the morning. Ella joined me in the room as I hung up, feeling somewhat better.

"I think I'm heading upstairs now, Anne," Ella said in her weary voice. "Beverly is ready too, so we'll see you in the morning. You should think about doing the same, Anne. It's been a long day."

"I know, and I will," I assured her. "I just want to make sure everyone will keep warm. Thank you so much for everything."

"I think in the morning we'll set pancakes in motion," she suggested. "We also have lots of sausage and plenty of Christmas stollen to offer."

"Sounds perfect, but as stuffed as I am right now, it doesn't sound so good," I responded holding my tummy. "I hope there's power for the coffee maker in case we have early risers."

"Good idea," she nodded. "Some may not sleep at all. By the way, your Uncle Jim is zonked out on the living room couch. I think he had a little more than his share of wine tonight."

"I noticed," I said shaking my head in disgust. "He's going to have one large headache in the morning, I'm afraid. They don't call that wine Red Devil for nothing!" Ella laughed.

As Ella kissed me good night on the cheek, Beverly came to say goodnight as well. I walked the two of them to the staircase, telling them to be careful and hoping that no one decided to walk in their sleep tonight. They laughed and held on to each other as they slowly climbed the long staircase.

When I walked back into the den, Devin and Sue were standing arm and arm trying to decide if they should turn in for the night. I had hoped I wasn't putting Sue in an awkward position by putting her and Devin in the same room for the night.

"Anne, the evening was spectacular," Devin said. "The dinner, the company and even the candlelight made it all very special."

"Thanks, I'm so glad you could join us," I responded. "Did you get the roll away in place with enough covers?"

"We're all set," Sue chimed in. "We just hope the little guys sleep through the night. Before I completely forget,

Anne, Mother and Dad said to wish you a Merry Christmas!"

"How nice," I smiled. "I meant to give them a call myself but just had too many distractions here! Is your Dad feeling okay?"

"I don't think he feels really great," she said sadly. "He never complains you know."

"We'll say goodnight for now unless you see us roam the halls. Are there any ghosts you need to tell us about?" teased Devin.

"Oh dear, Devin, you really don't want to go there," jumped in Sue as we both laughed. "Goodnight, Anne, and thanks again for a great evening."

We all hugged and off they went to the little maid's room.

I came back to the study and the few last survivors of the evening. Aunt Julia, Jack, and Sally were huddled by the fire as if they were afraid to leave it. It appeared they were having a jolly discussion from the laughs they were generating.

"Where is Sarah?" I asked Aunt Julia.

"She went on up an hour or so," Aunt Julia noted. "She got pretty bored with us. She's likely asleep by now. She was up very early this morning. Aren't you dead tired by now, Anne?"

"I haven't had much time to think about it," I said as I sat down next to Sally.

"How are you holding up?" I asked Sally.

"Oh, just fine," she shared. "I don't sleep well. I'll take the recliner in your room and read if you won't mind. I'm sure you're tired enough to fall asleep as soon as you hit that bed."

"Fine by me!" I responded. I wondered how Sam would feel about her sleeping in his chair. Why did I always feel Sam was watching everything?

"Jack, are you doing okay?" I asked since he was being so quiet.

"Very content, I must say," he said leaning back. "I'm still digesting that wonderful meal. The Red Devil did its part as well, so life is good." We all chuckled.

"You know, I just sipped a little with dinner, but everyone seemed to enjoy it," I confessed.

"Well, I think you need to slip off those darling shoes you've been wearing all night, put your feet up, and enjoy some of that wine," Aunt Julia suggested.

"I agree with her, Anne, but you'll have to do it without me," Sally said standing up. "I'm turning in, but I think you deserve a break today or let's say this morning!" I had to laugh as I kicked off my shoes.

"Follow me, Sally," Aunt Julia said. "We'll go up together. I'm afraid of the dark and that Grandmother ghost that occupies this house." Jack looked at me strangely.

"Oh, you party poopers," I called out to them. They walked out of the room.

# CHAPTER 72

I think they gave you some good advice, and I'm off to get you that glass of wine," Jack said going out of the room before I could respond. The idea sounded great. After all, everyone seemed to be settled for the night but me, and of course Jack. Just as I really did remove my shoes, Jack came into the room with my Red Devil.

"I think you made a big dent in that case of wine you purchased at the Wine Gala," Jack commented as he sat on the couch next to my chair.

"Indeed," I agreed. "I purchased another case to give away when folks leave. I'm not sure it will be as welcoming in the morning as it would have been last evening. Uncle Jim certainly enjoyed it!" Jack nodded and laughed.

"Yeah, he's in for a pretty unpleasant morning, I'm afraid," noted Jack. "You know, I'm a little puzzled about why the two of them got divorced. They seem to get along okay." I sighed.

"Uncle Jim went astray with an affair when he worked with Sam at Martingale's," I revealed. "Aunt Julia was devastated. She wrote him off completely after the divorce, but has now decided, for Sarah's sake, she needs to be civil to him."

"People make foolish mistakes throughout their lives," Jack said sadly. "I sense they really do still care for one another, or is it my imagination?"

"Yes, to some extent," I concurred. "Uncle Jim is very helpful to them at the shop, which I commend him for. He's dating some gal right now, but I think it's just to get Aunt Julia's attention."

"I am also curious about your friend Sally," Jack stated. "She started to tell me about losing someone special before we got interrupted."

"Yeah, that would be Tim," I recalled. "It's a long story, but in the end, she realized it wasn't ever going to go anywhere, and as a result, she was very hurt. She is not a social butterfly, so she may not meet anyone for some time. It's hard to move on when you've been hurt."

"Is that like anyone else I know?" he hinted.

I paused, thinking about how to answer.

"A break up is a little different than losing someone in death," I said defensively. "Someone intentionally hurting you is different than an unfortunate death."

"Of course, Anne, but hurt and pain come in different packages," Jack explained. "Do you see where I'm going with this?"

"I do, of course, but I don't feel that I have built a defensive wall around myself because I am hurting," I defended.

"That's good, because you have a lot of life ahead of you," Jack said as he took a sip of wine.

"So, Reverend Jack Fletcher, am I your normal widow from what you have seen in your profession?" I asked suggestively. He took a long look at me.

"You're not like anyone I have ever met, Anne Dickson," he firmly stated. "Even if I discount the fact that I am very attracted to you, I think you are anything but normal." Did I just hear him correctly? I turned to look away. "You are an over-achiever who has had a great deal of success in life. Most folks we consider normal, do not!" I decided to ignore his earlier comment.

"I know, I am very fortunate," I confessed. "I really don't know where my ambition and energy come from, but I have always been thankful for them. The down side is that not everyone can tolerate a woman like me. Sam and I both had ambitious goals and employees to manage so we understood each other immensely. We both needed each other to understand or our marriage would have never worked. I know that kind of relationship is rare! I cherish those years!" I had to hold back tears as I analyzed it all.

"As you should, Anne," he comforted. "Does this mean you have shut the door to sharing your life with anyone else?" The question threw me by surprise.

"In all fairness, Jack, I cannot answer that question," I said as politely as I could.

"That's a good sign, I think," Jack answered back. "Perhaps if someone found the right key to open one of the doors, you would respond. Did it make you feel uncomfortable when I told you I was attracted to you?" I smiled.

"I'm flattered, for sure. Of course not," I admitted as I blushed. "I picked up on your interest at the Wine Gala. I just didn't know how to react to it."

"Well, I have to admit, I am pretty savvy and sensitive to how people react to me because of my profession," he began to explain. "I think I detected you were attracted to me to some degree!" I was shocked at the confidence of this man! "Timing is everything in a relationship, I think. I respect who you are and trust that you will know when it is time to move on in your life."

"Thanks," I said not looking at him.

# CHAPTER 73

All of a sudden, on went the lights and the Christmas music that had been playing during dinner. The timing was good because I was ready to end this conversation.

"And there was light!" Jack joked as he held up hands. "How about I help you blow out those hundreds of candles. Hopefully, the lights will stay on."

"Good idea, Jack, but don't remove the candles in case we need them again," I added. "Jack, before you go; I just want you to know, I appreciate your understanding and I consider you a good friend." He looked at me as if he wanted to respond, but Uncle Jim came into the room.

"Well, hallelujah!" Uncle Jim yelled out. "We do have lights once again!" He looked terrible and quite shuffled. "Say, anyway we can get some coffee this early?" Jack and I looked at each other looking for the right words. I think we both wanted to laugh. "Hey, I didn't interrupt anything, did I?"

"I think coffee would be good for you and frankly, I could use a cup myself," I said, leaving the room.

The three of us went into the kitchen as Uncle Jim complained about his headache. I handed him some aspirin out of the kitchen cabinet. Once again Jack and I exchanged glances.

"Anne, why don't you go up for a while and try to get a few winks before everyone else wakes up," Jack suggested. "The coffee's going and I think everything's under control now with the lights and heat on."

"I think I will, Jack," I conceded. "I may not get any sleep, but I'd like to shower and change clothes before the breakfast crowd shows up."

"I'm going back to the couch," Uncle Jim said with exhaustion. "This room is moving. Jack, or should I say Reverend, would you let me know when the coffee is ready?" Jack nodded.

I smiled at them both, as I went up the stairs blowing out candles along the way. I stopped by Sammy's room where the night light showed his innocent face. He had missed a lot of the celebration, and it would be no time before he'd be wide awake.

I went into my bedroom and saw Sally leaning to one side of the recliner, sound asleep. She was even sawing a few logs as they say. I couldn't help but be pleased that she spent her holiday with her work family.

I changed into some jeans and a sweatshirt, which felt wonderful. I debated whether I would try to get a few winks of sleep with all the folks who may awaken at any given minute. When I walked over to my bedside, I noticed that Sam's photo was once again turned face down. Hopefully,

Sally didn't see this. I immediately turned it up knowing Grandmother wanted me to move on and ignore Sam.

As I sat there reminiscing about the evening, I was pleased that Jack and I had cleared the air on our feelings for one another. After all, I was still a grieving widow from a very happy marriage and he was the respectable reverend of my church.

I lay my head back on the pillow for just a moment so I could reflect on things a bit longer. My body was starting to give in and relax from the glass of wine I shared with Jack. My body then gave in to the rest it desperately needed. I was asleep in no time which had me dreaming about a journey with Santa Claus.

An hour or so later, I jumped at the sudden noise Sally made when she attempted to get out of the recliner.

"Sorry, Anne, please go back asleep," Sally insisted.

"What time is it?" I asked trying to regain my thoughts.

"It's only five o'clock, so try to get more sleep if you can," Sally suggested as she went into the bathroom.

"No, I'd better go downstairs to see if anyone is up," I said between yawns. "Ella will need help with breakfast." I turned to my bedside and noticed that once again Sam's photo was turned face down.

When Sally finished in the bathroom, I brushed my teeth and tried to make myself look presentable.

When I came down the stairs in the early morning darkness, I found my way to the kitchen where I saw Sue with Eli. He looked as if he was ready to start his day.

"Unfortunately, he's used to waking up around this time every day," Sue said looking very tired. "I didn't want him waking up the others in the room, so here I am. Thanks for making the coffee."

"We've had partakers throughout the night," I said pouring myself a cup. "I wonder if the snow plow has come through. I sure didn't hear it like I usually do."

"It didn't look like it had been here when I looked out the window," reported Sue. "I bet there's a foot of snow out there. You may be stuck with us for a while. Remember when we got snowed in at your mother's house one winter? We had people lying all over the place. What fun we had." We both laughed.

"It was wonderful," I recalled. "I'll never forget some of the stories we told or let's say announced that night!" Sue nodded and smiled.

"That reminds me of Muffin; she was with us that night," Sue added. "Oh, she is probably making a mess all over the house since we didn't come home last night."

"That's right. What about Uncle Jim's dog, Lucky?" I added. "Speaking of Uncle Jim, has he resurfaced this morning? He was up for a while last night."

"No, I haven't seen him," she said propping up her tired head. "It was such a unique evening, Anne. I peeked in the study and saw the reverend sleeping on the couch. I bet he'll remember this Christmas!"

"What do you honestly think of him, Sue?" I asked knowing she'd be very honest. She grinned.

"I should be asking you that question," Sue said grinning. "There's nothing not to like."

"He is really nice, but I'm just not ready to think of anyone at this point in my life," I confessed.

"Then let it be," she stated. "He's not going anywhere. I think if he saw the least bit of interest coming from you; he'd wait it out. Trust me."

"You give me too much credit, Sue," I responded. "He could have his pick of many, many single women who do not have a child. Hey, how did things work out in the maid's room last night?" She blushed.

"We did fine," she said laughing. "We were afraid to move so we wouldn't wake anyone. It got to be kind of silly. We started giggling and couldn't stop. The bed was shaking and we thought for sure someone would wake up." I had to laugh picturing the site.

"He's such a great guy, Sue," I said with sincerity.

"He is, but I also know what the challenges would be if this moves forward, so I'm giving this a lot of time. In the meantime, I'm sure having fun!"

# CHAPTER 74

Beverly joined us for a cup of coffee.

"I think I heard noises out in front of the house, Anne," Beverly reported. "Is anyone else up?"

"Let's go see," I said as the two of us went to open the front door.

There were Kip and Kevin shoveling snow off the front porch and steps in very little light.

"Hey, what's up here?" I asked trying to shield myself from the cold blowing in the house.

"I called, and the snow plow is in the neighborhood so we've got to start clearing sidewalks and paths so folks can get to their cars," Kip said out of breath. "The weather reports that all is clear for the rest of the day."

"Good news guys. Thanks so much. Please come in and get some hot breakfast and coffee when you can take a break," I encouraged. "We're doing pancakes and sausage."

"Sounds fantastic," yelled Kevin from the bottom step.

We shut the door and I told Beverly that we needed to start another pot of coffee and start frying sausage. She volunteered for the task as I tried to organize the kitchen table for the plates, cups, utensils, and slices of Christmas stollen. Sue took Eli with her to check on Devin and Mia.

"Well, look who is up and at 'em!" I said with surprise when Ella and Sammy walked into the kitchen. "Merry Christmas, my sugar plum!" I took him out of her arms to kiss and hold him tight.

"He's likely pretty hungry, sleeping through a long night," Ella thought. "That sausage sure smells good. I bet he could eat a little of that and maybe a piece of pancake when we get them going. I better get on that. You girls sure made a good start here and I'm ready for that cup of coffee!"

Jack now joined us in the kitchen and offered to help so we all had suggestions for him.

"Did you get any rest?" Jack asked me.

"I did doze off, and it helped immensely!" I beamed.

"I wish I had the proper clothes and shoes to help those guys out there," Jack offered. "They have got to be freezing."

"They wanted to clear places to walk because the snow plow is on its way." I explained. "Those guys are used to working in the outdoors so they have all the boots and clothing in that office. I think Kip said he thought we got 8 inches.

"Hey, Sammy, how are you today?" Jack asked going over to Sammy's chair.

Sammy grinned and immediately took his sippy cup and threw it on the floor. Jack laughed and picked it up. I took it from Sammy and once again told him no. I explained to

Jack it was a cute little trick of his that he liked to do when he was showing off. Jack joked that sometimes he brings out the worst in people and thought it was pretty cute.

"Hey, you all hear that noise out there?" Uncle Jim called out as he swaggered to the counter to get more coffee.

"How's your head?" asked Ella as she refilled his cup.

"It's calmer but feels bigger," he said as he put his hands on his temples. We all snickered, and I don't think he even noticed.

"I'm sure a bit of that fresh cold air out there would do your sinuses some good," Jack said jokingly.

"Very funny, Reverend," Uncle Jim answered.

"There will be some of these pancakes ready shortly, so why don't we give those guys a holler to come in," Ella suggested.

"Good idea, Ella," I said rushing to the front door to tell them.

I couldn't believe my eyes when I saw Abbey out there with them with a shovel in her hand. She looked hilarious in some of the guy's clothing and some big old boots.

"Hey, take a break and have some hot breakfast," I hollered into the cold air.

"Okay, good news," Abbey responded. "We'll all come into the sun porch and eat out there instead of tracking through the house."

"We'll be ready for you!" I yelled in return.

I went to the kitchen to tell Ella. She now had Jack doing nothing but flipping pancakes as she mixed more batter. He looked so intent on his task, I wanted to laugh. This was a fun and unique Kodak moment. We all helped to get their plates and coffee cups filled before they came in the door.

Mother and Harry now joined us looking rather chipper and well-rested. Harry said that he really liked the mattress on the bed, and Mother said that she sure was happy when that heat came back on again. I told them to help themselves and that there would be room on the sun porch if they liked.

"I heard Pat and Helen stirring when we came down," Mother reported.

"Any sound from Aunt Julia and Sarah's room?" I asked her.

"Not a peep," Mother said. "How did Sammy do during the night?"

"He slept like a lamb," I was proud to say.

Mother went over to give him a squeeze and to tell him how much he was missed last night.

"He sure has some unopened toys to play with today," noted Harry as he poured his own coffee.

"Auntie, Auntie," cried a little voice from Mia. She came running and hugged my legs.

"Hey pretty girl, would you like some pancakes?" I asked her as I leaned down to kiss her on the cheek.

"She loves pancakes," Sue quickly said in response holding Eli. "Devin is freshening up, and I'm warning you ahead of time that he is really starved. How can anyone be starved after that huge dinner we had last night?" I had to agree with her.

"Step right up, Miss Mia, and have a pancake," called out Jack. There was so much laughter and teasing in the kitchen, I had to chuckle to myself. Ella was giving Jack a hard time about his apron and that he was flipping faster and faster. I wouldn't have imagined how in the world this all could happen after such an elegant dinner party.

# CHAPTER 75

Devin, Sue, and the children started filling up the dining room table. Uncle Jim soon followed teasing the little ones. I don't think my dining room table would ever see two consecutive meals ever again. Beverly was going everywhere, filling up everyone's coffee cups. She had it in her blood to serve. I saw that in her the first time I met her.

Helen and Pat arrived in the kitchen and were amazed at the activity.

"How did you all sleep last night?" I asked as I gave them each a plate.

"We did fine," said Helen with a happy face. "How long have you been up, Anne?"

"Most of us were up between five and six o'clock to get the breakfast going," I answered. "Kip, Kevin, and Abbey were up earlier shoveling out front."

"I haven't had pancakes in a long, long time, but I suddenly feel the urge to have some," Helen bragged. "The heat sure feels good!"

"I agree, Mother," Pat chimed in. "The pancakes smell delicious. By the looks of Sammy, they must be darn good!" We all laughed as we looked at his self-feeding face.

"You won't get any kisses from anyone looking like that, Sammy boy!" I teased as I tried to wipe his face. They gave him a kiss anyway as they made their way to the pancakes.

Aunt Julia, Sarah, and Sally were the last to trickle down the stairs. They all were bragging amongst each other about the great night's sleep they had. I couldn't help but think that Sally's back had to be sore after sleeping in that recliner all night. I watched them fall in line as Jack filled their plates high with stacks of pancakes. There was no way he could leave that grill. I'm sure the task was way more than he bargained for. I watched everyone file in toward the dining room table as if they were at camp finding their spot.

The sound of the snow plow was loud enough for everyone to hear. Kip and Kevin immediately put on their coats to instruct them. Abbey was finished eating as well, and she decided to start collecting dirty plates for the dish washer. Poor Ella would likely be the last to leave the kitchen.

"Hey Abbey, how did you manage last night with the boys?" I teasingly asked.

"Oh, we had plenty of laughs, and I discovered a little refrigerator full of alcoholic refreshment," she shared with a laugh. "Did you know that was there, boss lady?"

"I knew it came from somewhere after some of those hot summer nights working in the heat outside," I revealed. "I certainly don't object when I see how hard they work, and I am going to pretend I didn't hear this!" Abbey laughed.

"Anne, I still can't believe you offered to finance my Christmas Shop," she brought up. "I couldn't think of anything else last night."

"You should have been thinking of your wedding announcement not the Christmas shop," I teased. "I'm so happy for the two of you and do hope that you will consider 333 Lincoln for the wedding. I don't think there's been a wedding here in its past."

"We'll see, but speaking of the past, how is that Taylor House book coming along?" Abbey said seriously.

"I am living it, Abbey!" I stated. "Life here is moving so quickly since Sam and I bought the place. I barely record anything lately. I do try to keep up with Sammy's baby book, which is very time consuming."

Pat had managed to take Sammy out of his chair and said they were going to take him in the study where his Christmas toys were. It was a great idea because we hadn't opened some of the gifts that were just for him. I told her I would follow after I got my camera.

Before I joined them, I took a picture of Jack flapping pancakes, Sue in her crazy work outfit, the stacks of dirty dishes, the folks still sitting at the dining room table, and now Sammy with his Grandmother Dickson and Auntie Pat.

Sammy knew he was the center of attention and loved playing the role as he tore paper and quickly tried out each new toy. Oh, if Sam could see this moment! Here was his son's first Christmas and he was putting on such a show. Now and then he wanted to take off crawling somewhere, but Sarah kept him in place. The room was filling up as everyone brought their coffee cups with them when they finished their breakfast. It was like another delayed Christmas party, and everyone was in a jolly mood.

"Oh, what a darling little tool box that is," I said to Sarah. "Who gave him that? I can tell after the tools go in his mouth, he will love exploring with that."

"The card said it was from Jack," she innocently revealed.

I looked at Jack who was standing by the door way observing Sammy. The smile on Jack's face said that he had given him a pretty cool gift. I waited until he looked my way before I acknowledged him with thumbs up. I had no idea he had brought him something. I took a photo of Sammy with his tool box and then went over to express my thanks. "

"I had a little toy box something like that when I was little," Jack confessed. "I played with that a long time. Every little boy should have his own tool box so he can pretend to fix things."

"You're right; there's plenty to be fixed around here," I teased. "He acts like he already knows what to do with it. Everything right now goes in the mouth since he's teething, of course." He laughed.

"These tools are multi-purpose," he added with a wink.

As I turned back to look at Sammy, I couldn't help but think how Sam loved his tools that were still in place in the garage. He loved tinkering around this place. It was one of the attractions when he fell in love with 333 Lincoln. He soon found out he had so little time, and others had to maintain this place. I wondered what he would be thinking now as he watched another man supply his little boy with his first tool box.

# CHAPTER 76

My cell went off in my pocket and saw that it was Kip calling. He said that all was clear enough if anyone needed to leave. I thanked him with all my heart before I hung up.

"Anne, I went into our website and there are some orders that have come in," Sally reported. "You may want to check for phone messages as well. Right now, I think we can get it all done tomorrow. Have you heard from Jean?"

"Thanks, Sally," I said taking a deep breath. How easy it was for me to get distracted with family and forget about shop obligations.

"I'll be happy to go straight there from here today," Sally offered.

"I think that would be a good idea, if you don't mind," I conceded. "I need to stay here with Pat and Helen until they fly out tomorrow, I'm afraid."

"Sure, as you should," Sally noted. "You have so much going on here. I won't open the shop, just check the orders and get ready for tomorrow."

"You're a gem," I told her. "Did you get plenty to eat?"

"I am stuffed to the max," she said holding her stomach. "I haven't had such good pancakes or any pancakes in a long time. I think that pancake flipper is a keeper, Anne."

"I'm not going there, Sally," I said as she left. She shook her head and smiled.

"I've asked Uncle Jim to drive us home this morning, Anne," Mother said with concern. "I don't want Harry driving in this weather."

"Good idea, but you are also welcome to spend the rest of the day if you'd like," I suggested.

"Oh my, no, we need our medications and a change of clothes," she said in defense.

"Well I hope he's sober enough," I added. "He's had a ton of coffee so hopefully will get you home safely. Kip said everything is clear to get the cars so we can call him to bring Uncle Jim's car around if he wants."

"Thanks, sweetheart," Mother said giving me a kiss. "We had a wonderful time, and it was good seeing Helen and Pat once again."

"That goes for me too," said Harry, giving me a big wet kiss on the cheek.

Uncle Jim walked over to me and said he already had the car out front. He was definitely sober from the cold air and promised he would get them home safely. I asked him to also walk them into the house as I doubted their sidewalks had been cleared.

"I don't know if we'll be able to leave here with all the new toys Eli and Mia are playing with," Sue excitedly explained. "It's Christmas all over again here with all of Sammy's toys."

"You all stay as long as you like," I encouraged. "We have plenty of leftovers if you're still here at lunch."

"Oh, no," she responded. "We all have eaten our fair share of pancakes to last a life time."

"They were the best ever, Anne. Thank you so much!" Devin added.

They both went back into the study and I went to the kitchen to check on Ella and Beverly who were doing clean up. As I looked about the room, it was indeed helpful to have Sam's great kitchen design that could accommodate big meals. He would have been proud to see us in action.

"Do you think we'll need to do a lunch, Anne?" Ella asked in her weary voice.

"I don't think so," I said laughing. "I think two big meals in a row should hold them."

"I'm surprised Julia and Sarah are still here," Ella stated. "Why didn't they get a ride with Jim?"

"Uncle Jim said he offered, but Aunt Julia's independent attitude said no thanks!" She laughed and shook her head. "I think everyone is having a good time and they just don't want to tackle the cold drive. I can't say I blame them."

As I walked back to join the others, I felt a hand on my shoulder. I turned around and it was Jack with his coat on.

"I think I need to be going, Anne, as much as I dread going out there," said Jack adjusting his woolen scarf. "It's a Christmas I will never forget. You really know how to entertain and make everyone feel comfortable. You know you could always turn this place in to a bed and breakfast. I think you know how to do it now." We laughed.

"Funny, but no thanks," I said as he got closer. "You could also get a job as a short order cook, I might add. No one could top your pancake skills from what I saw." He grinned.

"Funny, as well, Anne," he said with his best smile. "I think this is the beginning of a long friendship. You are a very special woman and mother. God made a gem when he made you." I blushed. "If I don't see you before the New Year begins, let me wish you a very Happy New Year!"

"You the same, Jack," I said as he suddenly kissed me on the side of my neck to my surprise.

"Be careful out there!" I said backing off.

He turned and went out the door. I looked out the window and saw Kip offer to get his car. How sweet and thoughtful that was of Kip to be this helpful. I watched Jack get in his car. I was really glad he came. He added to the mix of my family and friends and became very helpful in the process.

# CHAPTER 77

Sammy was still entertaining Helen, Pat, and Sarah. Sue, Devin, Mia, and Eli were having a little party of their own with some of the new toys. The room was out of control with laughter and screams. I tried to ignore them as I picked up wrappings and cups and saucers before I joined Ella and Beverly in the kitchen.

"Miss Anne, I better be going as well," announced Beverly. "I promised my Aunt Gertie I would stop by sometime today."

"I'm sending some leftovers with her," Ella said, holding a bag of containers. "She was more help than you can ever imagine. We still have enough to feed anyone that should be around later."

"I loved every minute of it," Beverly happily responded. "You have such a fun group here!"

"That we do," I said laughing at the noise in the background. "Don't forget to come into the shop this week so we can start training you. I don't think the boys will need you around here, but check with Kip first."

"Oh, I will," Beverly said with eagerness in her voice. "I can use the extra work right now."

"Thanks so much for the cute growth chart you gave Sammy," I noted. "I will put that up soon. I never thought of getting one of those before."

"I thought he'd have so many toys, and I was right by the sight of things!" she added. We laughed in agreement.

"He's having the time of his life right now," I said shaking my head in wonder.

"I have never seen him so wound up," Ella chimed in. "I hope he'll be able to settle in for a nap this afternoon."

"We'll see; it's his first Christmas you know!" I noted with a big smile on my face.

"I'll be going then," Beverly said putting on her coat. "Thanks again for everything as well as these yummy leftovers for later."

"It's our pleasure!" I said giving her a hug. "Happy New Year if we don't see each other before then."

After she left the kitchen, I told Ella she should take a break and put her feet up from the long day. I teased her that if she wasn't careful another meal would have to be served.

"Well, the next one will be a help yourself kind of deal," she stated. "We have a lot of beds to make up and still more dishes to wash before they get fed again!"

"We can always order pizza tonight, Ella," I suggested. "Helen seemed to enjoy that the last time they visited."

"That would be fine, too," Ella nodded. "She sure seems to be feeling pretty well since she arrived. I was worried about her at first."

"I know," I nodded. "I think she's had too many distractions to think about how she feels." Ella laughed in agreement.

When Ella and I joined the others in the study, Aunt Julia and Sarah were saying good-bye to everyone so I got their coats.

"Anne you totally outdid yourself with this one!" Aunt Julia bragged. "Your hospitality was over the top and we managed to get a good night's sleep as well. I think the colder temperatures were the trick. I told Sarah we're going to turn our heat down at night from now on."

"I'm so glad," I said giving her a hug. "Sarah you were such a big help with Sammy and a good sport about everything." She blushed.

"So, do you have New Year's Eve plans?" Aunt Julia asked.

"No, and that's fine by me," I stated. "I have a lot to catch up on at the shop from taking days off. When are you doing inventory?"

"Oh, don't bring that up," Aunt Julia said waving her hand. "Sarah is spending the night with a friend, so maybe I can think of something fun and clever for us. Thanks again for everything!"

As I walked them to the door, I realized that my house was going to be empty soon. I hoped everyone would make it home safely. When I opened the door, the frigid wind blew right into the house. I hated winter with a passion, and it was just beginning.

I went to get warm by the fire in the study, and as I observed more Christmas cheer, I couldn't help but feel a pang of sadness for the Forester family right now. I bet they couldn't wait for the holidays to be over. Tragedies on, or near, the holidays make everything more dramatic.

I heard the land line ring above the noise and went into the kitchen to answer.

"Jean, how good to hear from you!" I happily greeted her. "Did you have a good Christmas?"

"Jolly good, thanks! The frightful weather kept us close to the fire overnight by golly. The shop is shut today, right?"

"Right," I answered almost in her English brogue. "We need to open tomorrow so do you think Al could bring you in? Sally said we have orders waiting. By the way, the whole dinner party spent the night here last night. Sally just left here a while ago, and she was going by the shop."

"You don't say, Miss Anne," she said in disbelief. "I cannot wait to hear about each moment! I'll be there sharper than a butcher knife then!" We laughed.

I hung up remembering how much I missed her. She and Al would have really added to the party, if we could have kept him away from the alcohol. She was a vital part of my shop family. They had a routine of being with their neighbors that Al would not consider breaking. I was lucky to get them both to my Christmas party.

When I came out of the kitchen, Sue and Devin were bundling up the children to leave. It appeared they were taking off to the North Pole by the looks of them.

"I put the roll away bed back for you, Anne," Devin reported. "It sure came in handy. Spending a night in a maid's room was a first for me!" We had to chuckle.

I hated to see Mia and Eli go. They were turning out to be fun and loving cousins to my Sammy. Eli was starting to fuss as we lingered over the good-byes. Devin left us to go warm up the car for his temporary family.

"Plans for New Years?" I asked Sue.

"Yes, but if we don't find a sitter, Devin promised to cook a nice dinner for us," she said grinning.

"Mia and Eli like him very much, don't they?" I hinted. She smiled.

"For sure, but we'll see," she said as she lead her family to the front door. "Happy New Year, Anne, and thanks again for a wonderful couple of days!"

Suddenly the house was quiet. There were no more little voices. I didn't even hear Sammy's voice. I looked into the study and saw Sammy on Helen's lap as she was reading him The Three Bears. I grabbed my camera again for a precious Kodak moment. I hoped Sam was watching from above.

# CHAPTER 78

When I put Sammy down for a nap, I went into Helen and Pat's room where they were packing. Kevin was going to take them to the airport quite early in the morning.

"I'm glad you have Ella here to help you with everything," expressed Helen. "She really cares for Sammy, and he loves her too!'"

"No question," I agreed. "It was a win-win for both of us. I can't imagine arranging for a babysitter every time I go in and out of this house. I suppose I could have child care put into the flower shop." They chuckled.

"You still love your work, Anne?" Pat asked with curiosity.

"I really do, Pat," I answered with a big smile. "It's the sweet and bittersweet of your community. Having my shop located on historic Main Street is a bonus, for sure. I truly care what happens to the area and want to make it only better for Colebridge."

"I can see how you feel that way," voiced Pat. "I'm so glad we came and didn't miss all the excitement here in the last couple of days!" We all laughed. "Is there something we can do to help you get things back to order here?"

"Ella's got a system, and I just do what I'm told," I teased. "So much of the china has to be hand washed. That's what's she doing right now." I hesitated before bringing up the next topic. "I want you both to know that I'm planning on taking a trip this winter. Sam gave me a trip to London some Christmases ago and it just could never happen so I'm giving it serious consideration. I have an employee from there, and I would take her with me. She could show me around and I could meet her family in Bath which would be very nice." I watched to see the reactions on their face before I continued. "Please don't think I won't make that visit to come and see you both. In all honesty, I need to go somewhere different to clear my head a bit."

"You are so wise, honey," Helen finally said. "Sam would love for you to cash in on his thoughtfulness. It sounds wonderful!"

"Oh, Anne, it would be grand," Pat said with excitement. "Jane Austen's museum is in Bath which is the first thing I thought of. I know how much you appreciate her."

"Who would watch Sammy?" Helen asked with concern.

"Ella." I quickly noted. "Mother and Harry of course would help, and even Aunt Julia. I would only be gone ten days to two weeks, I suppose. I don't think I could be separated from Sammy any longer than that."

"Sounds like a wonderful plan, Anne," Pat surmised.

"Thank you both for all your support," I said trying not to tear up. "I love you both very much."

"We love you, too!" Helen said as she gave me a little hug.

The rest of the evening was very relaxing and spent getting 333 Lincoln back to order. Ella provided a light supper of leftovers which tasted just as delicious as they did the first time. Sammy continued to delight us all as we sat around the kitchen table laughing and sharing stories of the past twenty four hours.

We were all ready to turn in knowing everyone would be getting up early the next morning. I desperately needed sleep before my big day back to work. There was something wonderful about what would be normal again. A new year would begin shortly and there was much to look forward to.

# CHAPTER 79

The next morning was bright and sunny, but it was sad to say good-bye to Helen and Pat. I loved having a part of Sam near me for the holidays. Having them here for Sammy's first Christmas was important to all of us. I waved good-bye and slowly made my way back into the house to get warm. Ella was there looking at me as if she had experienced tremendous relief. I knew she, too, was looking for that sense of normalcy again.

I went upstairs to change into work clothes and was pleased to see Sam's photograph standing upright on my night table. I had to smile about that crazy Grandmother of mine. I was proud of how she had behaved throughout the holidays. For the first time, there was no appearance of lemonade, no surprise gifts, and no disappearances. I knew she wasn't really gone. She was just settling in here with the rest of us.

After I finished dressing, I opened my laptop to check my emails. Next to my computer were stacks of notes that I had been sporadically accumulating for The Taylor House book. I was so busy living the history that I didn't have time to record any of it. I sat down and started remembering how I thought the story would end up being about the love triangle of Marion Taylor, Albert Taylor, and my Grandmother Davis. Now I knew the house contained much more than the past.

With the purchase of the Brody property next door, the Dickson's were making an impact visually with the enterprise of Dickson Properties. There was a new generation growing up now in this magnificent home. It was my intention for Sammy to grow up here and enjoy the benefits that Sam and I had provided him. Sam had just a few years on this amazing hill in Colebridge, but he left his handiwork and a loving wife, who agreed to the purchase of 333 Lincoln because she fell in love with an adorable potting shed.

I told myself I would keep writing as I lived in this grand house with Sammy. The title of this book would have to change and be called the History of the Dickson House. I smiled as I closed my laptop. I would get back to this desk in the days ahead, feeling confident about the future.

I got in my car and headed back to the job I loved at Brown's Botanicals. As I drove the snow lined brick streets, I saw everyone get back to their normal work after Christmas. What a Christmas it was! We all had to move forward and use our history only as a resource for what was yet to come. Colebridge was a special place, and I was so proud to be part of it. It was a community that would always continue in many hearts and lives.

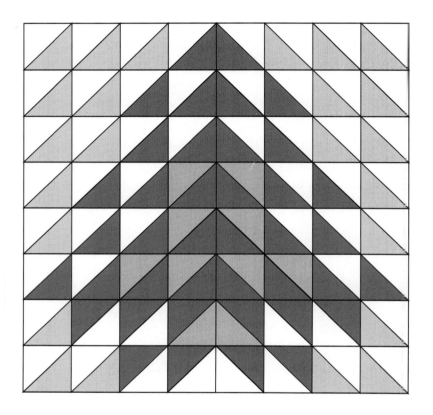

# The End

It is with mixed emotions that I say good-bye to the Colebridge community. I, like some of you, became emotionally attached to its characters and their daily lives.

It was a trip down memory lane, writing about things I love through the seasons of the year and my beloved Main Street. I also enjoyed sharing with you the trials and celebrations we all experience throughout our lives.

Thank you for encouraging me with your support and praise. This experience has made me a better writer and I look forward to taking you on another journey in the near future.

Ann Hazelwood

## JEAN'S ENGLISH SHORTBREAD

2 sticks of softened butter
2 cups of flour
½ cup confectioner's sugar
¼ teaspoon of salt.

Preheat oven to 325 degrees.
In a medium bowl, beat butter until creamy.
In another bowl, mix flour, sugar and salt.
Add dry mixture to creamed butter.
Stir well, until the consistency of dough.
Press dough into an ungreased 9X9 sized pan.
Use a fork to prick the dough all over, about 20 times.
Bake for 25 minutes or until done.
While the shortbread is still warm, cut into squares.
Enjoy with your favorite topping.

# ELLA'S CHRISTMAS WASSIAL

1 Gallon apple cider
1 Large can of pineapple juice
1 Cup of orange spice herb tea
1 Tablespoon whole cloves
1 Tablespoon whole allspice
2 Cinnamon sticks
Square of muslin cloth
Small piece of string

Mix the juices together in a big pot or crockery pot.
Put the spices in the middle of the
small square of muslin cloth and tie the
strong into a little bundle.
Put the spice bag in the pot and
let the whole thing simmer for 4-8 hours.
Enjoy!

# Cozy up with more quilting mysteries from Ann Hazelwood...

## WINE COUNTRY QUILT SERIES

After quitting her boring editing job, aspiring writer Lily Rosenthal isn't sure what to do next. Her two biggest joys in life are collecting antique quilts and frequenting the area's beautiful wine country. The murder of a friend results in Lily acquiring the inventory of a local antique store. Murder, quilts, and vineyards serve as the inspiration as Lily embarks on a journey filled with laughs, loss, and red-and-white quilts.

## THE DOOR COUNTY QUILT SERIES

Meet Claire Stewart, a new resident of Door County, Wisconsin. Claire is a watercolor quilt artist and joins a prestigious small quilting club when her best friend moves away. As she grows more comfortable after escaping a bad relationship, new ideas and surprises abound as friendships, quilting, and her love life all change for the better.

Want more? Visit us online at ctpub.com